T0418796

ADVANCES IN CANCER RESEARCH

Cancer Health Equity Research

VOLUME ONE HUNDRED AND FORTY SIX

ADVANCES IN CANCER RESEARCH

Cancer Health Equity Research

Edited by

MARVELLA E. FORD

Professor, Department of Public Health Sciences; Associate Director, Population Sciences and Cancer Disparities, Hollings Cancer Center, Medical University of South Carolina, Charleston; SmartState Endowed Chair in Cancer Disparities Research, South Carolina State University, Orangeburg, SC, United States

NESTOR F. ESNAOLA

Professor of Surgery, Houston Methodist Academic Institute/Weill Cornell Medical College; Division Chief of Surgical Oncology and Gastrointestinal Surgery, Department of Surgery, Houston Methodist Hospital; Surgical Director and Associate Director of Community Engagement and Cancer Control, Houston Methodist Cancer Center; Katz Investigator, Houston Methodist Research Institute, Houston, TX, United States

JUDITH D. SALLEY

Chair, Department of Biological & Physical Sciences, South Carolina State University, Orangeburg, SC, United States

Academic Press is an imprint of Elsevier
50 Hampshire Street, 5th Floor, Cambridge, MA 02139, United States
525 B Street, Suite 1650, San Diego, CA 92101, United States
The Boulevard, Langford Lane, Kidlington, Oxford OX5 1GB, United Kingdom
125 London Wall, London, EC2Y 5AS, United Kingdom

First edition 2020

ISBN: 978-0-12-820176-3
ISSN: 0065-230X

For information on all Academic Press publications
visit our website at https://www.elsevier.com/books-and-journals

Publisher: Zoe Kruze
Editorial Project Manager: Leticia M. Lima
Production Project Manager: Vijayaraj Purushothaman
Cover Designer: Miles Hitchen

Typeset by SPi Global, India

Contents

Contributors

Latecia M. Abraham
Department of Library Science and Informatics, Medical University of South Carolina, Charleston, SC, United States

Anthony J. Alberg
University of South Carolina, Arnold School of Public Health; Department of Epidemiology and Biostatistics, University of South Carolina, Columbia, SC, United States

Nusayba A. Bagegni
Washington University in St. Louis School of Medicine, St. Louis, MO, United States

Nadine J. Barrett
Duke Cancer Institute, Duke University Medical Center; Department of Family Medicine and Community Health; Duke Clinical and Translational Science Institute, Duke University School of Medicine, Durham, NC, United States

Colleen E. Bauza
Department of Public Health Sciences, Medical University of South Carolina, Charleston, SC, United States

Kenisha Bethea
Duke Clinical and Translational Science Institute, Duke University School of Medicine, Durham, NC, United States

Susan Bolick
South Carolina Department of Health and Environmental Control, Columbia, SC, United States

Heather S. Bonilha
Medical University of South Carolina, College of Health Professions, Charleston, SC, United States

Erika T. Brown
Morehouse School of Medicine, Atlanta, GA, United States

Marino A. Bruce
Program for Research on Faith and Health, Center for Research on Men's Health, Vanderbilt University, Nashville, TN, United States

Debbie C. Bryant
Department of Public Health Sciences, Clemson University, Clemson; Medical University of South Carolina, Hollings Cancer Center, Charleston, SC, United States

Dana R. Burshell
South Carolina Clinical & Translational Research Institute, Clinical and Translational Science Award, Medical University of South Carolina, Charleston, SC, United States

Clare E. Burton
Department of Pathology & Laboratory Medicine, Hollings Cancer Center, Medical University of South Carolina, Charleston, SC, United States

Kimberly Cannady
Academic Affairs Faculty, Medical University of South Carolina, Charleston, SC, United States

Kathleen B. Cartmell
Department of Public Health Sciences, Clemson University; Medical University of South Carolina, Hollings Cancer Center, Charleston, SC, United States

Courtney Chavis
Department of Public Health Sciences; Hollings Cancer Center, Medical University of South Carolina, Charleston, SC, United States

Maritza Chirinos
El Centro Hispano, Durham, NC, United States

Elise D. Cook
Department of Clinical Cancer Prevention, The University of Texas MD Anderson Cancer Center, Houston, TX, United States

Brittney Crawford
Hollings Cancer Center, Medical University of South Carolina, Charleston, SC, United States

Joan E. Cunningham
The National Coalition of Independent Scholars, San Antonio, TX, United States

Nestor F. Esnaola
Professor of Surgery, Houston Methodist Academic Institute/Weill Cornell Medical College; Division Chief of Surgical Oncology and Gastrointestinal Surgery, Department of Surgery, Houston Methodist Hospital; Surgical Director and Associate Director of Community Engagement and Cancer Control, Houston Methodist Cancer Center; Katz Investigator, Houston Methodist Research Institute, Houston, TX, United States

Victoria J. Findlay
Department of Pathology & Laboratory Medicine, Hollings Cancer Center, Medical University of South Carolina, Charleston, SC, United States

Laura J. Fish
Duke Cancer Institute, Duke University Medical Center, Durham, NC, United States

Marvella E. Ford
Professor, Department of Public Health Sciences, Medical University of South Carolina; Associate Director, Population Sciences and Cancer Disparities, Hollings Cancer Center, Medical University of South Carolina, Charleston; SmartState Endowed Chair in Cancer Disparities Research, South Carolina State University, Orangeburg, SC, United States

Starr Frazier
Department of Biological and Physical Sciences, South Carolina State University, Orangeburg, SC, United States

Antiqua Gathers
Department of Biological and Physical Sciences, South Carolina State University, Orangeburg, SC, United States

Kadeidre Gaymon
College of Nursing, Medical University of South Carolina, Charleston, SC, United States

Ronald L. Godbee
The River Church, Durham, NC, United States

Mathew J. Gregoski
Department of Arts & Sciences, Campbell University, Buies Creek, NC, United States

Derek M. Griffith
Center for Research on Men's Health; Center for Medicine, Health and Society, Vanderbilt University, Nashville, TN, United States

Demetrius Harvey
Black Men's Health Initiative, Wilson; Alumni Chapter of Kappa Alpha Psi Fraternity, Inc., Smithfield, NC, United States

Tonya R. Hazelton
College of Nursing, Medical University of South Carolina, Charleston, SC, United States

Ebony Hilton
Department of Anesthesiology and Perioperative Medicine, Medical University of South Carolina, Charleston, SC, United States

Daniel L. Howard
Public Policy Research Institute and Department of Sociology, Texas A&M University, College Station, TX, United States

Deborah Hurley
South Carolina Department of Health and Environmental Control, Columbia, SC, United States

Pao Hwa-Lin
Chinese Christian Church, Raleigh; Department of Medicine, Duke University School of Medicine, Durham, NC, United States

Kearston L. Ingraham
Duke Cancer Institute, Duke University Medical Center, Durham, NC, United States

Emily C. Jaeger
Center for Research on Men's Health, Vanderbilt University, Nashville, TN, United States

Kendrea D. Knight
Department of Public Health Sciences, Medical University of South Carolina, Charleston, SC, United States

Rita M. Kramer
Department of Hematology/Oncology, Medical University of South Carolina, Charleston, SC, United States

Bradley Krisanits
Department of Pathology & Laboratory Medicine, Hollings Cancer Center, Medical University of South Carolina, Charleston, SC, United States

Thomas A. LaVeist
Program for Research on Men's Health, Hopkins Center for Health Disparities Solutions, Johns Hopkins Bloomberg School of Public Health, Baltimore, MD; Department of Health Policy and Management, Tulane School of Public Health and Tropical Medicine, New Orleans, LA, United States

Claudia Lawton
Institute of Psychiatry, Medical University of South Carolina, Charleston, SC, United States

Peter Le
St. Joseph's Primary Care, Raleigh, NC, United States

John S. Luque
Institute of Public Health, College of Pharmacy and Pharmaceutical Sciences, Florida A&M University, Tallahassee, FL, United States

Franshawn Mack
Department of Biological & Physical Sciences, South Carolina State University, Orangeburg, SC, United States

Gayenell Magwood
Hollings Cancer Center; College of Nursing, Medical University of South Carolina, Charleston, SC, United States

Angela M. Malek
Department of Public Health Sciences, Medical University of South Carolina, Charleston, SC, United States

Erica Martino
Department of Public Health Sciences; Hollings Cancer Center, Medical University of South Carolina, Charleston, SC, United States

Leslie A. Moore
College of Medicine, Medical University of South Carolina, Charleston, SC, United States

Catishia Mosley
South Carolina Department of Health and Environmental Control, Columbia, SC, United States

Georges J. Nahhas
Hollings Cancer Center; Department of Psychiatry and Behavioral Science, Medical University of South Carolina, Charleston, SC, United States

Lynne H. Nguyen
Department of Health Disparities Research, The University of Texas MD Anderson Cancer Center, Houston, TX, United States

Lourdes Nogueira
Department of Pathology and Laboratory Medicine, Medical University of South Carolina, Charleston, SC, United States

Steven R. Patierno
Department of Medicine, Division of Medical Oncology, Duke University Medical Center, Durham, NC, United States

Lauren C. Peres
Department of Cancer Epidemiology, H. Lee Moffitt Cancer Center and Research Institute, Tampa, FL; Department of Epidemiology, Rollins School of Public Health, Emory University, Atlanta, GA, United States

Lindsay L. Peterson
Department of Medicine, Washington University in St. Louis; Washington University in St. Louis School of Medicine, St. Louis, MO, United States

Jaime F. Randise
Department of Pathology & Laboratory Medicine, Hollings Cancer Center, Medical University of South Carolina, Charleston, SC, United States

Schenita Randolph
Duke University School of Nursing, Durham, NC, United States

Judith D. Salley
Chair, Department of Biological & Physical Sciences, South Carolina State University, Orangeburg, SC, United States

Joellen M. Schildkraut
Department of Cancer Epidemiology, H. Lee Moffitt Cancer Center and Research Institute, Tampa, FL; Department of Epidemiology, Rollins School of Public Health, Emory University, Atlanta, GA, United States

Kit N. Simpson
Medical University of South Carolina, College of Health Professions, Charleston, SC, United States

Roland J. Thorpe Jr.
Program for Research on Men's Health, Hopkins Center for Health Disparities Solutions; Department of Health, Behavior & Society, Johns Hopkins Bloomberg School of Public Health, Baltimore, MD, United States

David P. Turner
Department of Pathology & Laboratory Medicine, Hollings Cancer Center, Medical University of South Carolina, Charleston, SC, United States

Keith E. Whitfield
Provost and Senior Vice President for Academic Affairs, Wayne State University, Detroit, MI, United States

Ping Zhang
Chinese American Friendly Association, Raleigh, NC, United States

Preface

According to the National Cancer Institute, cancer disparities are defined as the differences in cancer outcomes across population groups. These outcomes include the following (National Cancer Institute, 2020):

- Incidence (new cases)
- Prevalence (all existing cases)
- Mortality (deaths)
- Morbidity (cancer-related health complications)
- Survivorship, including quality of life after cancer treatment
- Burden of cancer or related health conditions
- Screening rates
- Stage at diagnosis

This volume is the second cancer disparities focused volume of *Advances in Cancer Research*. Cancer disparities are a major public health problem in the United States (USA). Although cancer mortality rates are decreasing for the general population, the mortality rates are not decreasing as quickly for specific population groups such as African Americans and other diverse populations, rural populations, and the medically underserved. (Within the context of the volume, the terms "African American" and "black" are used interchangeably.) The cancer incidence and mortality rates that drive these disparities are the result of a complex interplay between the social determinants of health and biological factors. The goal of cancer disparities research is to ultimately reduce and eliminate these disparities.

Ten chapters in the volume highlight one or more cancer disparities outcomes. The chapters present information that readers could employ to design and implement interventions to reduce cancer disparities and promote cancer health equity. The content of the chapters includes five types of cancer for which significant disparities in mortality outcomes exist: ovarian, breast, prostate, and cervical cancer. Several of the chapters describe factors related to successfully launching and sustaining community outreach and engagement interventions to improve cancer health outcomes. The volume concludes with a primer on defining a catchment area, the first step toward developing and implementing these interventions. The synergy among the chapters is shown by their emphasis on the multiple factors that contribute to cancer disparities, ranging from social factors to biological factors.

1. Ovarian cancer

Chapter 1 by Peres and Schildkraut, "Racial/ethnic disparities in ovarian cancer research," presents epidemiological data showing racial/ethnic differences in risk and survival of ovarian cancer in the United States, with African Americans having the poorest outcomes. The multi-site African American Cancer Epidemiology Study (AACES), led by Dr. Schildkraut et al., was conducted from 2010 to 2015, and is currently the largest population-based case–control study of African American women with epithelial ovarian cancer, enrolling approximately 601 cases and 752 controls. The authors point to the need for additional large-scale studies with adequate sample size and greater representation of African American women. Future studies could lead to the identification and evaluation of environmental, genetic, and clinical factors associated with these disparities. The information that is gained from these future studies could lead to effective interventions to reduce ovarian cancer risk and improve survival due to this disease.

2. Breast cancer

Chapter 2 by Bagegni and Peterson, "Age-related disparities in older women with breast cancer," highlights the important area of age-related disparities in breast cancer mortality rates. This is a timely topic, considering the current increase in the proportion of older adults in the United States. Older women have a worse prognosis than younger women for early stage and advanced breast cancer, and many older women are undertreated. The authors note that undertreatment may contribute to age-related differences in breast cancer mortality, and recommend that treatment interventions become individualized, taking into account functional status as well as chronological age. These future age-tailored interventions could help to improve breast cancer survival rates among older women.

Chapter 3 by Krisanits et al., "Pubertal mammary development as a "susceptibility window" for breast cancer disparity," focuses on the biological mechanisms that potentially contribute to cancer disparities. Specifically, the authors evaluate the relationship between a high-fat Western diet and changes in pubertal mammary gland development that increase the risk of breast cancer. Given the racial disparities in obesity rates related to diet, pubertal mammary development is a particular risk factor for women from diverse populations. The proposed solution to reducing the diet-related breast cancer risk factors, and improving breast cancer prognosis, is to develop pre-/postnatal and childhood dietary lifestyle interventions to reduce fat intake, thereby lowering the age of menarche and pubertal mammary gland development.

Chapter 4 by Ford et al., "BMI, physical activity, and breast cancer subtype in white, black, and Sea Island breast cancer survivors," includes a population group that is unique to the coastal southeastern United States; the African American Sea Island/Gullah population. The Sea Islanders are one of the most racially homogenous group of blacks in the United States and represent an understudied group in terms of cancer research. In their study ($n = 137$; 21% Sea Islander), Ford et al. note that the breast cancer risk/prognostic factors of high body mass index and low physical activity levels were higher in non-Sea Island African American women than in Sea Island African American women or in non-Hispanic white women. The frequency of estrogen receptor-negative breast cancer, with the poorest prognosis, was also the highest in the non-Sea Island African American women. Future studies with larger sample sizes could evaluate the impact of lifestyle interventions on reducing disparities as well as the potential impact of biological factors among these different racial/ethnic groups.

3. Prostate cancer

Chapter 5 by Thorpe et al., "Race differences in mobility status among prostate cancer survivors: The role of socioeconomic status," focuses on prostate cancer. The central question of their research is whether racial differences exist in mobility limitation among prostate cancer survivors, and whether socioeconomic status/position plays a role in this relationship. Their sample of 661 prostate cancer survivors (296 black men and 365 white men) was drawn from the Diagnosis and Decisions in Prostate Cancer Treatment Outcomes (DAD) Study. The findings show that after adjusting for age, marital status, health insurance, tumor differentiation and grade, treatment, and time to treatment, the African American men in the sample had a higher prevalence of mobility limitation in comparison to their white counterparts. This study highlights the importance of taking socioeconomic status/position into account when evaluating racial differences in functional outcomes related to prostate cancer.

4. Cervical cancer

Chapter 6 by Ford et al., "Assessing an intervention to increase knowledge related to cervical cancer and the HPV vaccine," brings to light the disparities in US cervical cancer mortality rates, which could be ameliorated by increased uptake of the HPV vaccine. The authors describe their evidence-based cancer education intervention focused on improving knowledge of cervical cancer, human papillomavirus (HPV), and the HPV vaccine among diverse populations in South Carolina. The results show that in the predominantly African American, medically underserved, and rural sample ($n = 276$; 93% African American), the intervention had a positive effect on increasing knowledge related to cervical cancer and the HPV vaccine. While knowledge is a necessary but insufficient element of health behavior change, subsequent increases in HPV vaccination rates in South Carolina will be monitored to assess the intervention impact. Future studies could include the educational intervention as part of a behavioral change intervention related to HPV vaccination, particularly in communities with the highest cervical cancer mortality rates.

5. Patient navigation

Chapter 7 by Cartmell et al., "Patient barriers to cancer clinical trial participation and navigator activities to assist," presents data from their feasibility study of patient navigation in lung and esophageal cancer clinical trial enrollment. This approach is described as a potential means of combatting the age, race, and gender disparities in US cancer clinical trial participation. The study includes 40 participants, 10 (25%) of whom are African American. The results show that the most frequently cited barriers to trial participation were fear, communication issues with healthcare professionals, with health insurance coverage, and transportation issues. About 50% of navigated patients had one or more stated barriers to care, although most of the reported barriers were concentrated in a small number of participants. Approximately half of the patients had no specific barriers to care and 10% reported 4–7 barriers. This study highlights the importance of integrating the navigator role and responsibility within the trial participants' clinical care team in future research.

6. Community outreach and engagement

Chapter 8 by Barrett et al., "Project PLACE: Enhancing community and academic partnerships to describe and address health disparities" links cancer disparities with the past lack of engagement of key community stakeholders into cancer prevention and control initiatives. The authors note that to successfully achieve cancer health equity, it is first necessary to engage community members in the design and implementation of these initiatives and interventions. Barrett et al. share their experiences and lessons learned in Project Population Level Approaches to Cancer Elimination (Project PLACE) at the Duke Cancer Institute in North Carolina. Their efforts have led to a National Cancer Institute-funded community health assessment that successfully reached 2315 participants over a 7-month period. The goal of the assessment is to develop future initiatives, with community input, to reduce cancer disparities and enhance cancer health equity in North Carolina and beyond. Project PLACE serves as a model for the design and conduct of future cancer research projects.

Chapter 9 by Griffith and Jaeger, "Mighty men: A faith-based weight loss intervention to reduce cancer risk in African American men," sheds light on the need for cancer prevention and control interventions for African American men. This is an important topic; in the United States, disparities in cancer mortality rates are driven by the high death rates among black men. There is a critical need to develop effective interventions to help prevent cancer and improve the cancer survival rates in this population. Griffith and Jaeger note that there are currently no evidence-based cancer prevention and control interventions related to nutrition and physical activity that have been developed specifically for African American men. The authors describe the conceptual framework, aims, and setting for their evidence-based intervention titled: *Mighty Men: A Faith-Based Weight Loss Intervention for African American Men*. The information shared could be used as a foundation for future cancer research interventions designed to reduce risk and improve cancer survival outcomes among black men.

Chapter 10 by Nguyen and Cook, "A primer for cancer research programs on defining and evaluating the catchment area and evaluating minority clinical trials recruitment," provides a guide for defining the catchment areas of National Cancer Institute-designated cancer centers. This is a very important and timely topic. Due to a recent change in funding requirements, cancer centers seeking NCI designation (or renewal) must now provide an accurate and meaningful description of their catchment areas. The definition of the catchment areas is linked to each center's ability to compile and interpret cancer health statistics for the catchment area, and to develop interventions to improve cancer health outcomes for the catchment area population. In their primer, Nguyen and Cook present a detailed description related to the proper definition of the catchment area and provide a case example from the University of Texas MD Anderson Cancer Center to illustrate best practice. This important and timely information could inform the practices of population sciences leaders at national and international cancer centers.

7. Summary

In summary, the chapters in this volume provide important information for the conceptualization, development, implementation, and evaluation of interventions to reduce cancer disparities and improve cancer health equity in diverse population groups. Despite the prevalence of cancer disparities, exciting opportunities exist to expand the scope of cancer disparities research and to reduce and even eliminate differences in cancer risk and prognosis.

Marvella E. Ford, Nestor F. Esnaola, and Judith D. Salley

Reference

National Cancer Institute. (2020) What are cancer disparitites. (https://www.cancer.gov/about-cancer/understanding/disparities).

CHAPTER ONE

Racial/ethnic disparities in ovarian cancer research

Lauren C. Peres*, Joellen M. Schildkraut

Department of Cancer Epidemiology, H. Lee Moffitt Cancer Center and Research Institute, Tampa, FL, United States

Department of Epidemiology, Rollins School of Public Health, Emory University, Atlanta, GA, United States

*Corresponding author: e-mail address: lauren.peres@moffitt.org

Contents

Abstract

Ovarian cancer is one of the most fatal cancers diagnosed in women in the United States (U.S.). Data from national databases, including the Surveillance Epidemiology and End Results (SEER) program, show racial/ethnic differences in risk and survival of epithelial ovarian cancer with higher incidence among white women yet worse survival among African-American women compared to other racial/ethnic groups. The reasons for these differences are not well understood, but are likely multi-factorial. Epidemiologic studies suggest there may be some risk factor differences across racial/ethnic groups that would explain differences in the incidence of this rare and heterogeneous disease. Likewise, although data suggest that socioeconomic factors and access to care contribute to the disparity in ovarian cancer survival among African-American women, there are likely other contributing factors that have not as of yet been identified. Small sample sizes of minority women from individual studies do not provide adequate power to evaluate fully the contributions of environmental, genetic, and clinical factors associated with ovarian cancer risk and survival within these groups. Pooling existing data from individual epidemiologic studies has made

Advances in Cancer Research, Volume 146
ISSN 0065-230X
https://doi.org/10.1016/bs.acr.2020.01.002

a valuable contribution; however, new data collection is warranted to further our understanding of the underpinnings of the disparities in ovarian cancer that may lead to prevention and improved survival across all racial/ethnic groups.

1. Introduction

Ovarian cancer is a rare and fatal disease that is also heterogeneous. The most common histologic group is epithelial ovarian cancer (EOC) that comprises ~90% of all ovarian cancer diagnoses (Clarke & Gilks, 2011) and is herein the focus of this chapter. With most studies of ovarian cancer largely consisting of white women, racial disparities in ovarian cancer risk and survival have been understudied. Only in the past decade has there been a serious effort to evaluate ovarian cancer risk and survival among other racial/ethnic groups. This is largely facilitated by the formation of consortia which have pooled data from existing epidemiologic studies providing larger sample sizes of minority women than would be available from any single study (Cannioto, Trabert, Poole, & Schildkraut, 2017). Along with data from large national databases, the Ovarian Cancer Association Consortium (OCAC) (Berchuck, Schildraut, Pearce, Chenevix-Trench, & Pharoah, 2008) and the multi-site African American Cancer Epidemiology Study (AACES) (Schildkraut et al., 2014) contribute most of the cross racial/ethnic group evaluations of risk and survival factors for ovarian cancer in the published literature. In the text below, we refer to "blacks" and "African Americans" interchangeably and similarly to "whites," "non-Hispanic whites (NHW)" and "European Americans."

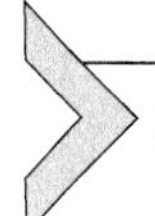

2. Disparities in ovarian cancer incidence rates in the United States

Ovarian cancer is a rare malignancy, accounting for approximately 3% of all cancer diagnoses among women, and it is estimated that 22,440 women will develop ovarian cancer in the United States (U.S.) in 2017 (Siegel, Miller, & Jemal, 2017). According to data from the Surveillance, Epidemiology, and End Results (SEER) program, the overall age-adjusted incidence rate of ovarian cancer is 11.7 cases per 100,000 women (Howlader et al., 2017). EOC incidence rates vary substantially across racial/ethnic groups. As shown in Fig. 1, NHW women have the highest incidence of EOC (12.4 cases per 100,000), followed by Hispanics (all races) at 10.6 cases

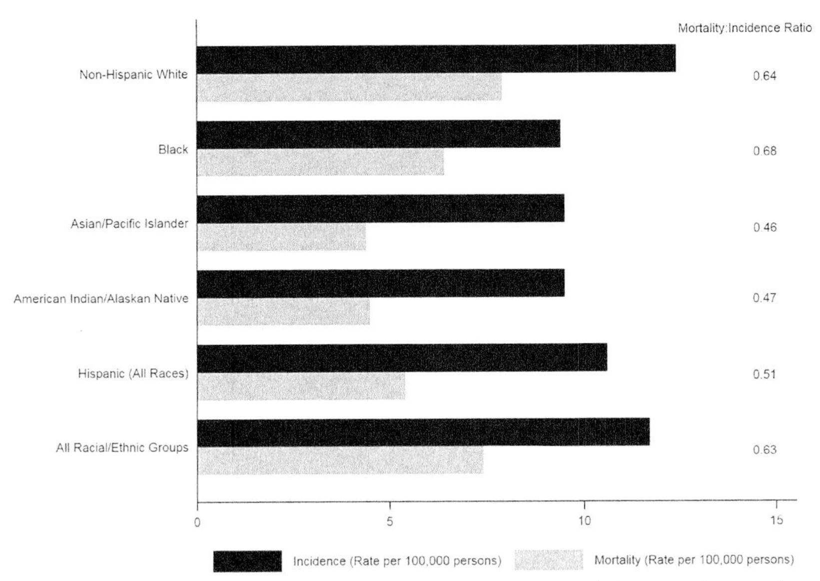

Fig. 1 SEER age-adjusted incidence rates, mortality rates, and the mortality-incidence ratio for invasive epithelial ovarian cancer by race/ethnicity, 2010–2014, SEER 18 registries. Incidence and mortality rates are age-adjusted to the 2000 U.S. Standard population (19 age groups = Census P25–1130). SEER 18 registries include San Francisco, California (CA); Connecticut; Detroit, Michigan; Hawaii; Iowa; New Mexico; Seattle, Washington; Utah; Atlanta, Georgia (GA); San Jose-Monterey, CA; Los Angeles, CA; the Alaska Native Registry; Rural Georgia; CA excluding San Francisco, San Jose-Monterey and Los Angeles; Kentucky; Louisiana; New Jersey; and Georgia excluding Atlanta and Rural Georgia. *Adapted using data from the SEER Cancer Statistics Review, 1975–2014 Howlader, N., Noone, A., Krapcho, M., Miller, D., Bishop, K., Kosary, C., et al. (2017). SEER cancer statistics review, 1975–2014. Bethesda, MD: National Cancer Institute, based on November 2016 SEER data submission, posted to the SEER web site, April 2017. Retrieved June 22, 2017, from http://seer.cancer.gov/csr/1975_2014/.*

per 100,000 women. Asian/Pacific Islanders, American Indian/Alaska Natives, and black women have the lowest incidence rates (9.5, 9.5, and 9.4 cases per 100,000 women, respectively) (Howlader et al., 2017).

The incidence of EOC overall has declined considerably over time, with an average annual percent change (APC) of −1.8% during 2005–2014 (Howlader et al., 2017). This has been largely attributed to an increase in the use of oral contraceptives over time, which confers a protective effect on ovarian cancer risk; however, other factors such as increasing fertility rates after World War II and a decline in menopausal hormone therapy after 2002 may also play a role (Gnagy, Ming, Devesa, Hartge, & Whittemore, 2000; Webb, Green, & Jordan, 2017; Yang et al., 2013). The rate of decline in incidence over time has also differed markedly by race/ethnicity, where the greatest decline was observed among white women in comparison to other races (−1.9% APC from 2005 to 2014 for NHW women while only −0.5% APC for other races) (Howlader et al., 2017). More recent trends also indicate that an even steeper decline in incidence is occurring for white women as of 2010, where the APC was −3.1% during 2010–2014 (Howlader et al., 2017).

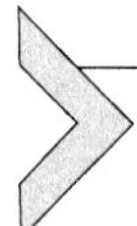

3. Disparities in ovarian cancer mortality rates in the United States

Ovarian cancer is the deadliest malignancy among all of the female reproductive system neoplasms. An estimated 14,080 women are expected to die from this disease in the U.S. in 2017 (Siegel et al., 2017) and an overall mortality rate of 7.4 deaths per 100,000 women was observed for 2010–2014 (Fig. 1) (Howlader et al., 2017). Similar to EOC incidence, NHW women have the highest mortality rate, 7.9 deaths per 100,000 women (Howlader et al., 2017). The second highest mortality rate is experienced by black women (6.4 deaths per 100,000 women), followed by Hispanics (all races) at 5.4 deaths per 100,000 women, and American Indian/Alaska Natives and Asian/Pacific Islanders with the lowest mortality rates (4.5 and 4.4 deaths per 100,000 women) (Howlader et al., 2017). Fig. 1 also shows the ratio of mortality to incidence by race/ethnicity, which provides an alternative way to assess disease burden that accounts for the effects of incidence on mortality. Black and NHW women have the highest mortality-to-incidence ratio, yet the ratio is slightly higher for black women (0.68). The mortality-to-incidence ratio for the other racial/ethnic groups is substantially lower, ranging from 0.46–0.51, with the lowest ratio for Asian/Pacific Islanders (0.46).

Across all racial/ethnic groups, the mortality rate for ovarian cancer has declined 2.2% per year during 2005–2014 (Howlader et al., 2017) and is largely a consequence of the decline in ovarian cancer incidence (Sopik, Iqbal, Rosen, & Narod, 2015). NHW women experienced the largest decrease in mortality during this time period (−2.4% APC) in comparison to the other racial/ethnic groups. The rate of decline for black women was −1.1%, while the APC was <1% for Hispanic, American Indian/Alaska Native, and Asian/Pacific Islander women (−0.7%, −0.7%, and −0.5%, respectively).

4. Disparities in ovarian cancer survival in the United States

Survival for EOC is poor, with data from SEER suggesting that only 46% of women with EOC in the U.S. survive 5 years after diagnosis (Siegel et al., 2017). Considerable variation in EOC survival is observed by race/ethnicity, with African-American women experiencing the poorest 5-year relative survival (36%) while survival for white women is approximately 46% (Siegel et al., 2017). Data from the Southwest Oncology Group (Albain, Unger, Crowley, Coltman, & Hershman, 2009), which includes 1429 advanced-stage ovarian cancer patients treated on randomized phase III clinical trials from 1974 to 2001, similarly observed a 10-year survival rate of 13% for African-American women versus 17% for all other patients, and increased mortality for African-American women in comparison to all other patients (hazard ratio [HR] = 1.61, 95% CI = 1.18–2.18). Sparse data exist on the survival of other, less common racial/ethnic groups, yet there is evidence to support a better survival among Hispanic and Asian women in comparison to white and African-American women (Fuh et al., 2015; Ibeanu & Díaz-Montes, 2013; Park, Ruterbusch, & Cote, 2017). Ibeanu and Diaz-Montes evaluated survival by ethnicity using SEER data and observed that Hispanic women had a higher median survival in comparison to NHW and non-Hispanic black (NHB) women (45 months compared to 36 and 24 months, respectively) (Ibeanu & Díaz-Montes, 2013). Fuh et al. evaluated differences in survival between white and Asian women in SEER from 1988 to 2009 (Fuh et al., 2015). The 5-year disease-specific survival for Asian women was considerably higher than white women, 59.1% and 47.3%, respectively. Even after adjustment for age and year of diagnosis, surgery, stage and grade, Asian women still had significantly better outcomes than white women. The authors also divided Asian women into ethnic subgroups and observed that Chinese, Korean, Filipino, and Vietnamese women had better survival in comparison to Japanese and Indian/Pakistani women, which remained statistically significant in multivariate analyses. A more recent analysis of SEER data (2000–2009) (Park et al., 2017) observed similar survival patterns, where Hispanic and Asian women had similar or improved outcomes in comparison to NHWs, while NHB women had the poorest survival.

Due to the sparse data evaluating racial/ethnic survival patterns for ovarian cancer, we conducted an analysis of SEER data which compared the overall survival of women diagnosed with EOC during 2004–2014 by race/ethnicity. To complete these analyses, data were restricted to the main epithelial histotypes as defined by the 2014 World Health Organization Classification of Tumors of Female Reproductive Organs (Kurman, Carcangiu, Herrington, & Young, 2014) (high-grade serous, low-grade serous, endometrioid, clear cell, mucinous, carcinosarcoma, and malignant Brenner tumors), and women with available follow-up information (exclusion of women diagnosed upon autopsy or death certificate) as well as known race/ethnicity and disease stage. In Fig. 2, Kaplan-Meier survival curves are provided stratified by the extent of disease (localized/regional, distant stage) due to the substantially different survival outcomes for these groups of women. In comparison to other racial/ethnic groups, black women have markedly worse survival irrespective of disease stage yet more pronounced for localized/regional disease, and Asian/Pacific Islanders are experiencing the best survival outcomes.

Considerable differences in survival patterns over time are evident by race/ethnicity. Using SEER data from 1973 to 2007, Terplan et al. evaluated whether racial differences in ovarian cancer survival have increased over time (Terplan, Schluterman, McNamara, Tracy, & Temkin, 2012). Overall, African-American women had a higher hazard of all-cause mortality in comparison to white women, even after adjustment for important clinical and demographic characteristics (HR = 1.31, 95% CI = 1.26–1.37). The addition of surgical treatment to the multivariable models decreased the survival disparity slightly (HR = 1.27, 95% CI = 1.21–1.34), suggesting that a small portion of these differences are due to the receipt of treatment. The authors also observed that this survival disparity increased over time ($P < 0.01$), where the HR comparing African-American women to white women in 1973–1977 was 1.20 (95% CI = 1.06–1.36), but 1.65 (95% CI = 1.46–1.86) in 2003–2007. Similarly, a recent analysis of SEER data (Zeng et al., 2015) quantified whether survival has improved from 1990 to 2009 by age, race, and sex for multiple cancer types. In multivariate analyses adjusting for marital status, histology, SEER registry, stage, and age, ovarian cancer-specific survival significantly improved over time for white women, yet no change in survival was observed for Asian women. In contrast, African-American women experienced a significant decline in survival during the 20-year time period, where in comparison to 1990–1994, the HR for 2000–2004 was 1.32 (95% CI = 1.11–1.56) and for 2005–2009, the HR was 1.21 (95% CI = 1.02–1.45).

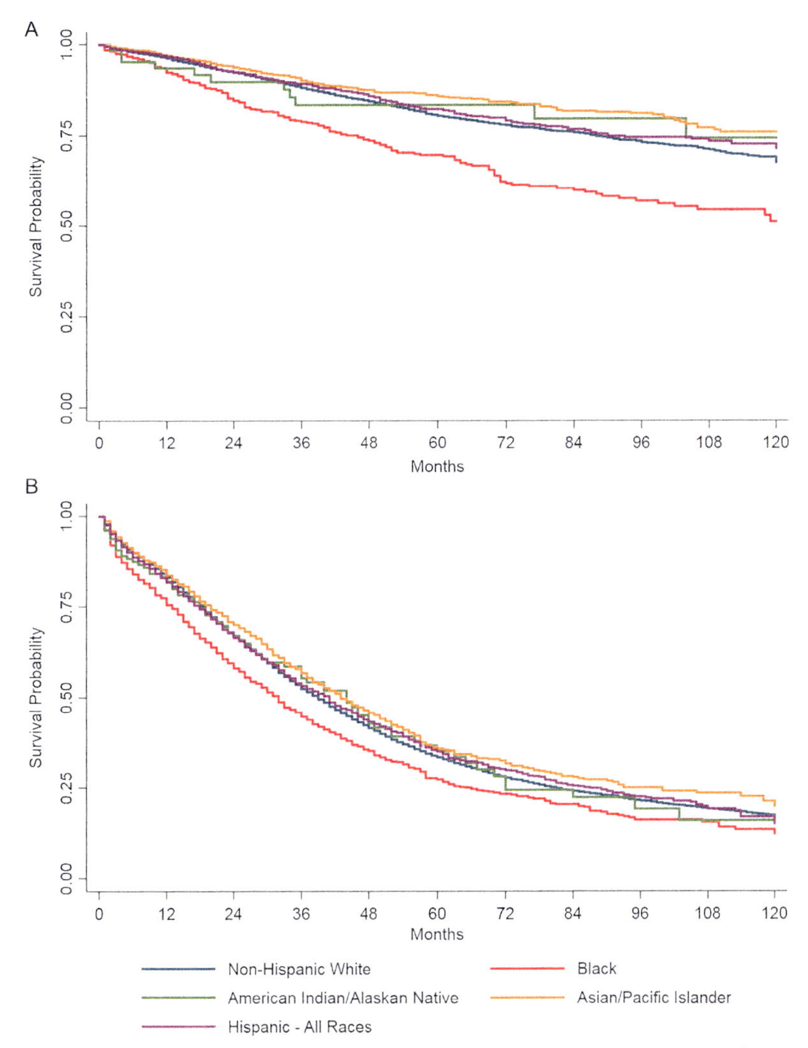

Fig. 2 Kaplan-Meier survival curves of invasive epithelial ovarian cancer by race/ethnicity and stage, 2004–2014, SEER 18 registries. (A) Localized/regional stage. (B) Distant stage. Data are restricted to the main epithelial ovarian cancer histotypes (high-grade serous, low-grade serous, endometrioid, clear cell, mucinous, carcinosarcoma, and malignant Brenner tumors) which have follow-up information (excludes cases diagnosed through autopsy or death certificate) and known race/ethnicity and disease stage. SEER 18 registries include San Francisco, California (CA); Connecticut; Detroit, Michigan; Hawaii; Iowa; New Mexico; Seattle, Washington; Utah; Atlanta, Georgia (GA); San Jose-Monterey, CA; Los Angeles, CA; the Alaska Native Registry; Rural GA; CA excluding San Francisco, San Jose-Monterey and Los Angeles; Kentucky; Louisiana; New Jersey; and GA excluding Atlanta and Rural Georgia.

5. Differences in tumor characteristics by race/ethnicity

Racial/ethnic differences in the occurrence of EOC stage and histologic subtype at diagnosis have been documented in the U.S. population and are likely to partially explain some of the disparity in incidence and survival across these groups. Racial/ethnic differences by stage at diagnosis are mainly observed in Asian women, where Asian women are more likely to be diagnosed with early stage disease compared to other racial/ethnic groups. A report by Fuh et al. (2015) used data from SEER (1988–2009) to evaluate differences in tumor characteristics among Asian (defined as Vietnamese, Filipino, Chinese, Korean, Japanese, and Asian Indian/Pakistani ethnicities) and white women with EOC and reported that 41.8% of Asian women were diagnosed with early stage disease (Stage I or II) compared to 30.3% of white women. Similarly, after excluding women with un-staged cancers, a more recent report of SEER data (2000 – 2013) observed that 36% of Asian women were diagnosed with localized or regional stage disease in contrast to <30% for NHW, NHB, and Hispanic women (Park et al., 2017). This finding is likely explained by the fact that Asian women are diagnosed more frequently with clear cell carcinoma (as discussed in detail below), which is predominantly an early stage disease (Peres, Cushing-Haugen, et al., 2019). There is some indication that black women may be slightly more likely to be diagnosed with late stage disease compared to other racial/ethnic groups (Beckmeyer-Borowko et al., 2016); however, this has not been consistently shown across studies (Park et al., 2017).

The most notable differences in the proportion of histologic subtype diagnoses by race/ethnicity are for clear cell and serous carcinoma. The most recent SEER analysis of racial/ethnic differences in EOC histology noted that Asian women were considerably more likely to be diagnosed with clear cell ovarian cancer (11.7%) than other racial/ethnic groups (ranging from 2.4% to 4.5%), with NHB women the least likely to be diagnosed with clear cell ovarian cancer (2.4%) (Park et al., 2017). In this analysis, which combined epithelial ovarian and fallopian tube cancer, Park et al. found serous cases to be the least commonly diagnosed among Asian women while NHB and NHW women had a comparable proportion of serous cancer diagnoses. Hispanic women fell in-between with a greater proportion of serous cancer diagnoses than Asians but less than NHB and NHW women. Another analysis of SEER data that focused on Asian and white women with ovarian cancer diagnosed in 1988–2009 (Fuh et al., 2015) showed that Asian women were twice as likely to be diagnosed with clear cell ovarian cancer (13.4%) compared to white women (5.6%) while serous EOC was less common among Asians (44.6%) compared to whites (56.5%). A study using data from the National Cancer Database evaluated histologic subtype differences by race/ethnicity stratified by disease stage, noting that NHB women were significantly more likely to be diagnosed with late-stage high-grade serous, clear cell, and mucinous carcinomas compared to NHW women (Beckmeyer-Borowko et al., 2016). The majority of previously published data on histologic subtype differences by race/ethnicity do not differentiate high versus low grade cancer diagnoses by histology which may impact the findings of the racial/ethnic group comparisons particularly for serous and endometrioid cancers. Nonetheless, the differences in histologic subtype distribution by race, with the consideration that treatment regimens are not uniformly effective across EOC histotypes (Brown & Frumovitz, 2014; Hess et al., 2004; Pectasides, Pectasides, Psyrri, & Economopoulos, 2006; Shida, Takabe, Kapitonov, Milstein, & Spiegel, 2008; Takano, Tsuda, & Sugiyama, 2012), may contribute to survival differences across racial/ethnic groups. Differences in incidence by histotype likely reflect differences in the prevalence of ovarian cancer risk factors (as described in Section 6) and may also have a genetic susceptibility component (as described in Section 7) that varies across racial/ethnic groups.

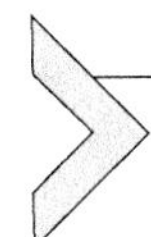

6. Epidemiologic characteristics contributing to ovarian cancer etiology

The prevalence of epidemiologic characteristics suspected or confirmed to play a role in ovarian cancer etiology has been shown to vary across racial/ethnic groups, and may partially explain the EOC incidence differences by race/ethnicity. For example, it is well established that nulliparity increases the risk of ovarian cancer, and that with each pregnancy, a woman's risk of ovarian cancer is reduced by ~10% (Wentzensen et al., 2016). Data from the National Center for Health Statistics shows that Hispanic and black women have a greater number of children in comparison to white women (Martin, Hamilton, Osterman, Driscoll, & Mathews, 2017). Another reproductive exposure, oral contraceptive use, is a well-established protective factor for EOC risk, and data from the 2011–2013 National Survey of Family Growth showed that among women aged 15–44 years, NHW women were more likely to currently use contraceptive pills than NHB women, 19.0% versus 9.8%, respectively

(Daniels, Daugherty, & Jones, 2014). It is possible that the differences in prevalence of these protective reproductive factors among African-American women is partly responsible for the lower EOC incidence rate in this racial/ethnic group.

At present, race- or ethnicity-specific risk factor associations have been compared in five studies (John, Whittemore, Harris, & Itnyre, 1993; Moorman, Palmieri, Akushevich, Berchuck, & Schildkraut, 2009; Ness, Grisso, Klapper, & Vergona, 2000; Peres, Risch, et al., 2018; Wu, Pearce, Tseng, & Pike, 2015), yet most are not inclusive of all racial/ethnic groups and include <150 cases of women with a non-white race/ethnicity. Moorman, et al. used data from a population-based case-control study in North Carolina to assess variation in the relative contributions of ovarian cancer risk factors between African-American and white women (Moorman et al., 2009). The majority of risk factor associations were similar across race/ethnicity, except for tubal ligation and family history of breast or ovarian cancer which were more pronounced among African-American women. However, the authors only adjusted for age in their models which may result in residual confounding and the sample size of African-American women was rather small ($n = 111$ cases), limiting their power to detect differences by race.

Using data from four ovarian cancer case-control studies in Los Angeles County, California, Wu, et al. evaluated the population attributable risk (PAR) for six ovarian cancer risk factors (parity, oral contraceptive use, tubal ligation, endometriosis, family history of ovarian cancer, and talc use) as a whole and separately by race/ethnicity (Wu et al., 2015). Significant differences in the prevalence of these risk factors by race/ethnicity were evident, most notably for nulliparity where the prevalence was higher among NHW women (nulliparous: 23.7% NHWs, 13.7% Hispanics, 16.8% African-Americans) whereas a higher prevalence of tubal ligation and talc use was observed in African-American women (tubal ligation: 14.1% NHWs, 26.3% Hispanics, 30.8% African-Americans; talc use ≥ 1 year: 30.4% NHWs, 28.9% Hispanics, 44.1% African-Americans). However, these racial/ethnic differences did not account for widely different risk factor associations, and similarly the PAR estimates for all six risk factors together were comparable, 57.9% (95% CI = 48.7–65.3) in NHWs, 56.1% (95% CI = 46.8–63.3) in Hispanics, and 53.8% in African-Americans (95% CI = 45.0–60.7). Also, there were no substantial racial/ethnic differences in the individual PAR% estimates for each risk factor separately. Considering the differences in oophorectomy rates by race/ethnicity in the general population, Wu, et al. also corrected the population at risk in the PAR for oophorectomy rates, yet little changes were made to the overall conclusion—only a third of the incidence differences by race/ethnicity can be explained by non-genetic ovarian cancer risk factors and oophorectomy rates. While this study provides some of the only data on Hispanic women, their results were constrained by sample size ($n = 128$ African-American and $n = 308$ Hispanic cases).

A recent manuscript (Peres, Risch, et al., 2018) improved upon the small numbers of non-white women in previous studies by capitalizing on pooled data from 11 case-control studies in the OCAC (Berchuck et al., 2008) and the largest population-based case-control study of African-American women with ovarian cancer, the AACES (Schildkraut et al., 2014). Risk factor associations for 17 established or suspected exposures involved in EOC etiology were estimated for NHW, Hispanic (all races), black and Asian/Pacific Islander women, and racial/ethnic heterogeneity was assessed by cross-product interaction terms. After correction for multiple comparisons, the only risk factor association which was significantly different across racial/ethnic groups was hysterectomy (false discovery rate P-value = 0.008), where a more pronounced positive association was observed among black women (OR = 1.64, 95% CI = 1.34–2.02). However, the direction of this association is controversial. A meta-analysis (Jordan et al., 2013) revealed that an inverse association for hysterectomy and EOC risk was observed for studies conducted prior to 2000, yet a positive association was observed for studies conducted after 2000, suggesting that a temporal shift may have occurred in this association. A study using AACES data (Peres et al., 2017) found that this shift may be partly due to changes in hormone therapy usage patterns over time. Nevertheless, the findings from Peres et al. provide additional support in a large pooled dataset that risk factor differences by race/ethnicity do not completely account for the racial/ethnic variation in EOC incidence (Peres, Risch, et al., 2018).

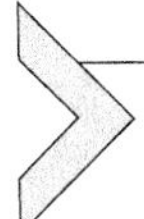

7. Genetic and molecular epidemiology of ovarian cancer

Several reports which have examined ovarian cancer risk and survival have concluded that environmental factors likely do not completely explain racial/ethnic differences in the incidence and survival of EOC (Bandera, Lee, Rodriguez-Rodriguez, Powell, & Kushi, 2016; Bristow et al., 2015; Farley et al., 2009; John et al., 1993; Moorman et al., 2009;

Ness et al., 2000; Peres, Risch, et al., 2018; Song et al., 2009; Wu et al., 2015). Therefore, genetic susceptibility has been hypothesized to play a role. Both genetic and molecular characterization of ovarian tumors has been vastly understudied in diverse racial/ethnic populations and published findings have often been in small and underpowered studies. Although efforts to better characterize genetic and genomic characteristics within racial/ethnic groups are currently underway, these efforts are at least a year away from completion. Below we summarize what is known to date.

7.1 Genetic susceptibility and mutations

7.1.1 Highly penetrant susceptibility genes

Being a mutation carrier for either of the breast-ovarian cancer susceptibility genes, *BRCA1* and *BRCA2*, has important implications related to screening and prevention of breast and ovarian cancers. Though rare, deleterious mutations in the *BRCA1/BRCA2* exhibit a Mendelian dominant mode of inheritance with incomplete penetrance and are associated with a high risk of ovarian cancer, often with an earlier age at diagnosis (Antoniou et al., 2003). There is a greater probability of developing ovarian cancer among *BRCA1* carriers (~40%) compared to *BRCA2* mutation carriers (11–18%) (Hall et al., 2009). Data on the prevalence of carriers in the general population is sparse especially for non-white subjects. Based on available data, the prevalence of *BRCA1* and *BRCA2* mutations are similar across racial/ethnic groups, from 1% to 4% in whites, Africans, Asians, and Hispanics in the general population (Kurian, 2010). An exception is among Ashkenazi Jews who have an approximately 10-fold higher prevalence of *BRCA1* and *BRCA2* mutations compared to the general U.S. population (Hall et al., 2009).

In cross-sectional analyses of a clinical database of women diagnosed with ovarian cancer, the prevalence of a *BRCA1* or *BRCA2* mutation was found to be ~7% of white European women, ~5% among Latin-American women, 5% among Asian women and 1% among African-American women (Hall et al., 2009). In a 2010 review by Kurian (2010), a much higher prevalence of *BRCA1* mutations was found among those with a family history of both breast and ovarian cancer; 25–40% of whites had a known deleterious mutation, while 16% of African-Americans, 12% of Asians, and 16–23% of Hispanics had deleterious *BRCA1* mutations. Among these women, less prevalent were mutations in *BRCA2* with a prevalence of 13% among Asian Americans, a range of 6–15% in whites, 11% in African-Americans, and a prevalence in Hispanics ranging from 6% to 8% (Kurian, 2010). The prevalence of variants of uncertain significance was highest among African-Americans (16–44%) and lower among Asians (14%), whites (5–12%) and Hispanics (9–23%) (Kurian, 2010). Aside from the higher prevalence among Ashkenazi Jews, the generally low prevalence of *BRCA1/2* mutations and the improved 5-year survival among carriers (Bolton et al., 2012) would support that mutations in *BRCA1/2* are unlikely to be a major factor in racial/ethnic disparities related to the incidence and survival of ovarian cancer.

7.1.2 Common variants

In comparison to rare deleterious genetic mutations, common genetic variants, defined as having a prevalence of $\geq$5%, typically classified as single nucleotide polymorphisms (SNPs), are more common but less penetrant and often associated with weak increased risk of ovarian cancer as demonstrated in the recent, large genome wide association studies (GWAS) performed almost exclusively among women of European ancestry (Phelan et al., 2017). Thus far, there have been 30 identified GWAS SNPs associated with ovarian cancer risk in women of European ancestry (Phelan et al., 2017). Of note, although the majority of GWAS associated SNPs are associated with all histotypes or the most common serous histotype, some are associated with less common subtypes. Five loci have been associated only with the mucinous histotype and one SNP was found to be associated only with clear cell EOC. In a recent report in which the contribution of the genome-wide associated variants was evaluated with respect to ovarian cancer risk, it was shown that the susceptibility variants identified thus far account for only a small portion of EOC risk (Clyde et al., 2016).

To date, there has been only one relatively small, published GWAS of EOC among Asian women from China, where two genome-wide significant SNP associations were reported (Chen et al., 2014) that appear unique to this Asian population when compared to >30 reported thus far among women of European ancestry (Phelan et al., 2017). Additional GWAS investigations are currently underway in other racial/ethnic groups. However, achieving an adequately powered sample size for a GWAS analysis is a considerable challenge, particularly among women of African ancestry.

7.2 Gene expression subtypes of high-grade serous ovarian cancer

A recent focus of pathologic-epidemiologic research is the determination of gene expression patterns to define distinct molecular subtypes of EOC mostly among high-grade serous ovarian cancer (Konecny et al., 2014;

The Cancer Genome Atlas Research Network, 2011; Tothill et al., 2008). However, this molecular characterization is not as advanced as with breast cancer subtypes where targeted treatments have been shown to be effective (Prat et al., 2015; Reis-Filho & Pusztai, 2011). It is hoped that better characterization will also lead to refinements in risk factor associations and the unmasking of causal pathways.

Most of the characterization of gene expression subtypes have been exclusively among white women with EOC and have not addressed women of diverse racial/ethnic and socioeconomic backgrounds. Only one report examined epidemiologic associations with molecular signatures in high-grade serous ovarian cancer (Schildkraut et al., 2013). The findings in this report suggest differences in the occurrence of molecular signatures among African-Americans compared to mostly white subjects (Tothill et al., 2008). The high stromal cluster, as described by Tothill et al. (2008), which is associated with more favorable survival, was found to be inversely associated with African-American EOC cases compared to non-African-American cases. Those with the mesenchymal/undifferentiated cluster, were found to be associated with worse survival and were more likely to be found among African Americans. It is possible that differences in the distribution of gene expression subtypes across racial/ethnic groups may contribute to differences in their survival. The higher prevalence of the aggressive triple negative breast cancer subtype among African-American women (Bauer, Brown, Cress, Parise, & Caggiano, 2007), that is in part responsible for worse survival compared to white women, serves as a prototypical example of possible implications of differences in cancer subtype distributions across racial/ethnic groups. To date, there are no studies that have adequately addressed ovarian cancer gene expression subtypes in non-white patients (Doherty et al., 2017). With ongoing studies in this area, more insight into the relationships between subtype and other patient characteristics among racial/ethnic groups is expected in the next few years.

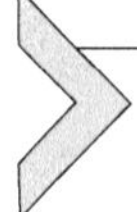

8. Disparities in treatment and access to care by race/ethnicity

Racial, ethnic and socioeconomic disparities with regards to access to care and treatment for ovarian cancer has been documented in several reports from population-based databases that support disparities in diagnosis and treatment among African Americans and women of lower socioeconomic status (SES) (Bandera et al., 2016; Bristow et al., 2015, 2013;

Peterson et al., 2014; Terplan et al., 2012). The recommended standard treatment for ovarian cancer includes cytoreductive surgery and a combination regimen of platinum- and taxane-based chemotherapy. In a recent report by Bristow et al., adherence to treatment guidelines from the National Comprehensive Cancer Network (NCCN) was evaluated in over 10,000 ovarian cancer patients with stage IIIC–IV epithelial ovarian cancer identified from the SEER-Medicare database (1992–2009) (Bristow et al., 2015). Analyses included data from various racial/ethnic groups: whites, blacks, Asian or Pacific Islanders, and Hispanics. The results indicated that black race and low SES was associated with non-adherent NCCN guideline care. Further analyses showed that SES and deviation from treatment guidelines were found to be associated with worse mortality and likely explains some of the observed survival disparity across racial, ethnic and socioeconomic groups. Similarly, Warren et al. used the National Cancer Institute's Patterns of Care data to evaluate whether receipt of guideline care and cancer mortality varied across race among ovarian cancer patients (Warren et al., 2017). The findings suggested that receipt of guideline care was low among all ovarian cancer patients, but more so for black women and this association could not be explained by provider characteristics or other factors. Consistent with this finding, a report that examined neighborhood SES using addresses for 581 ovarian cancer cases residing in Cook County, Illinois, including the city of Chicago, found that greater disadvantage was associated with higher grade tumors and suboptimal debulking as well as later stage at diagnosis, supporting the association of low SES with more advanced and aggressive disease (Peterson et al., 2014).

Even when access to care is not a predominant factor, racial differences in ovarian cancer survival have been observed. A possible explanation is found in analyses of data from Kaiser Permanente Northern California (KPNC), which showed that African-American women were more likely to receive chemotherapy dose reduction and experience the worst survival than any other racial/ethnic groups even after adjusting for clinical characteristics and detailed treatment information (Bandera et al., 2016). In this example, all women received health care and insurance coverage from KPNC, suggesting that the observed racial/ethnic differences may be due to other factors besides access to care. In a separate study of 1392 European-American and 97 African-American advanced stage Gynecologic Oncology Group trial participants, 5-year progression-free survival among women who were optimally debulked was 40% lower for African-American versus European-American women (Farley et al., 2009).

This is a noteworthy finding given the unique features of clinical trials such as stringent inclusion/exclusion criteria, receipt of uniform treatment, and the ability to control for debulking status and residual disease, which are major predictors of survival.

Treatment decisions and efficacy of treatment modalities can be impacted by comorbid conditions. An example of this is the variation in dosing of chemotherapies among patients who are overweight or obese. There is concern among these patients that if dosing were determined based on actual weight, treatment toxicities may arise. A recent study using data from Kaiser Permanente reported that obese women were more likely to receive lower doses of chemotherapy by body weight and dose reduction, which was associated with a worse survival (Bandera, Lee, Rodriguez-Rodriguez, Powell, & Kushi, 2015). There are known differences in the prevalence of obesity by race/ethnicity, with data from the National Health and Nutrition Examination Survey (NHANES) showing that among adults aged >20 years, 56.9% of NHB women were obese (BMI of $\geq 30\,kg/m^2$) compared to 45.7% of Hispanic, 35.5% of NHW, and 11.9% for non-Hispanic Asian women (Arroyo-Johnson & Mincey, 2016). Similarly, other comorbid conditions, such as diabetes and hypertension, are more prevalent among NHB and Hispanic women compared to white and Asian women (Benjamin et al., 2017). It is possible that some of the racial/ethnic differences in outcomes may be attributed to the differential burden of these comorbidities across racial/ethnic groups.

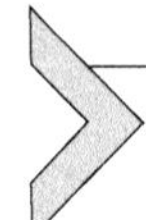

9. Challenges in studying racial/ethnic disparities in ovarian cancer

The study of racial/ethnic disparities in ovarian cancer has been hindered by the lack of racial/ethnic representation in existing epidemiologic studies. This is exacerbated by the rarity of this disease, making it difficult to acquire a sufficient population of non-White women. Furthermore, the histotype-specific risk factor and genetic SNP associations underscore the heterogeneity of ovarian cancer and the need for large sample sizes to characterize differences in risk factors and genetic susceptibility across racial/ethnic groups. Therefore, a multi-site/consortia effort is often required to perform adequately powered analyses within these non-white subgroups.

Until recently, the largest number of African-American cases enrolled in a single epidemiologic study of ovarian cancer was ~150 women.

However, considerable strides in the study of African-American women with ovarian cancer have been made with AACES (Schildkraut et al., 2014), the basis for many risk factor evaluations in African-Americans on which we report. AACES was conducted from 2010 to 2015 and is currently the largest population-based case-control study of African-American women with EOC, enrolling approximately 601 cases and 752 controls where race was determined by self-report. AACES was conducted in 10 geographic locations, representing a paradigm for conducting epidemiologic studies of rare cancers in minority populations. An alarming finding in the recruitment of women in AACES was that 15% of potential study subjects were deceased at the time of ascertainment, although a rapid-case ascertainment approach was utilized. This high proportion of African-American women with rapidly fatal cancer is in contrast to the experience of a similar population-based case-control study in North Carolina of mostly white women where only 4% of potential cases were deceased at the time of ascertainment (Schildkraut et al., 2014). This observation, along with the published reports we have summarized above, highlight the survival disparity in African-American women and the need for further study. Besides NHW women and some advancement in the study of NHB women, studies on other racial/ethnic groups are lacking.

The available reports on racial/ethnic groups are mainly limited to data from large national cancer databases (e.g., SEER) or population-based case control studies, where medical record data is often incomplete. In contrast, clinical trials often include detailed medical record information, yet do not typically include diverse racial/ethnic groups. Thus, it has been difficult to fully address whether treatment and access to care are driving the racial/ethnic EOC survival disparities. To provide insight on why specific racial/ethnic subgroups experience poorer outcomes for ovarian cancer, it is imperative that future clinical trials are more inclusive of all racial/ethnic subgroups and that clinic-based epidemiologic studies are conducted in order to obtain more detailed clinical data on a more racially heterogeneous population of women with ovarian cancer.

We conclude that new studies, involving both population-based and hospital-based approaches, are necessary to obtain a more complete assessment of disparities in ovarian cancer. Both study designs require a greater concerted effort to enroll minority women who have been underrepresented in existing studies, and new approaches may be needed to address gaps in the study of ovarian cancer disparities.

References

Albain, K. S., Unger, J. M., Crowley, J. J., Coltman, C. A., & Hershman, D. L. (2009). Racial disparities in cancer survival among randomized clinical trials patients of the southwest oncology group. *Journal of the National Cancer Institute, 101*(14), 984–992. https://doi.org/10.1093/jnci/djp175.

Antoniou, A., Pharoah, P. D. P., Narod, S., Risch, H. A., Eyfjord, J. E., Hopper, J. L., et al. (2003). Average risks of breast and ovarian cancer associated with BRCA1 or BRCA2 mutations detected in case Series unselected for family history: A combined analysis of 22 studies. *The American Journal of Human Genetics, 72*(5), 1117–1130. https://doi.org/10.1086/375033.

Arroyo-Johnson, C., & Mincey, K. D. (2016). Obesity epidemiology trends by race/ethnicity, gender, and education: National Health Interview Survey, 1997-2012. *Gastroenterology Clinics of North America, 45*(4), 571–579. https://doi.org/10.1016/j.gtc.2016.07.012.

Bandera, E. V., Lee, V. S., Rodriguez-Rodriguez, L., Powell, C. B., & Kushi, L. H. (2015). Impact of chemotherapy dosing on ovarian cancer survival according to body mass index. *JAMA Oncology, 1*(6), 737–745. https://doi.org/10.1001/jamaoncol.2015.1796.

Bandera, E. V., Lee, V. S., Rodriguez-Rodriguez, L., Powell, C. B., & Kushi, L. H. (2016). Racial/ethnic disparities in ovarian cancer treatment and survival. *Clinical Cancer Research, 22*(23), 5909–5914. https://doi.org/10.1158/1078-0432.CCR-16-1119.

Bauer, K. R., Brown, M., Cress, R. D., Parise, C. A., & Caggiano, V. (2007). Descriptive analysis of estrogen receptor (ER)-negative, progesterone receptor (PR)-negative, and HER2-negative invasive breast cancer, the so-called triple-negative phenotype: A population-based study from the California cancer registry. *Cancer, 109*(9), 1721–1728. https://doi.org/10.1002/cncr.22618.

Beckmeyer-Borowko, A. B., Peterson, C. E., Brewer, K. C., Otoo, M. A., Davis, F. G., Hoskins, K. F., et al. (2016). The effect of time on racial differences in epithelial ovarian cancer (OVCA) diagnosis stage, overall and by histologic subtypes: A study of the National Cancer Database. *Cancer Causes & Control, 27*(10), 1261–1271. https://doi.org/10.1007/s10552-016-0806-6.

Benjamin, E. J., Blaha, M. J., Chiuve, S. E., Cushman, M., Das, S. R., Deo, R., et al. (2017). Heart disease and stroke statistics-2017 update: A report from the American Heart Association. *Circulation, 135*(10), e146–e603. https://doi.org/10.1161/CIR.0000000000000485.

Berchuck, A., Schildraut, J. M., Pearce, C. L., Chenevix-Trench, G., & Pharoah, P. D. (2008). Role of genetic polymorphisms in ovarian cancer susceptibility: Development of an international ovarian cancer association consortium. *Advances in Experimental Medicine and Biology, 622*, 53–67.

Bolton, K. L., Chenevix-Trench, G., Goh, C., Sadetzki, S., Ramus, S. J., Karlan, B. Y., et al. (2012). Association between BRCA1 and BRCA2 mutations and survival in women with invasive epithelial ovarian cancer. *JAMA, 307*(4), 382–390. https://doi.org/10.1001/jama.2012.20.

Bristow, R. E., Chang, J., Ziogas, A., Campos, B., Chavez, L. R., & Anton-Culver, H. (2015). Sociodemographic disparities in advanced ovarian cancer survival and adherence to treatment guidelines. *Obstetrics & Gynecology, 125*(4), 833–842.

Bristow, R. E., Powell, M. A., Al-Hammadi, N., Chen, L., Miller, J. P., Roland, P. Y., et al. (2013). Disparities in ovarian cancer care quality and survival according to race and socioeconomic status. *Journal of the National Cancer Institute, 105*(11), 823–832. https://doi.org/10.1093/jnci/djt065.

Brown, J., & Frumovitz, M. (2014). Mucinous tumors of the ovary: Current thoughts on diagnosis and management. *Current Oncology Reports, 16*(6), 389.

Cannioto, R. A., Trabert, B., Poole, E. M., & Schildkraut, J. M. (2017). Ovarian cancer epidemiology in the era of collaborative team science. *Cancer Causes and Control, 28*(5), 487–495.

Chen, K., Ma, H., Li, L., Zang, R., Wang, C., Song, F., et al. (2014). Genome-wide association study identifies new susceptibility loci for epithelial ovarian cancer in Han Chinese women. *Nature Communications, 5*, 4682. https://doi.org/10.1038/ncomms5682.

Clarke, B. A., & Gilks, B. (2011). Ovarian carcinoma: Recent developments in classification of tumour histological subtype. *Canadian Journal of Pathology, 3*, 33–42.

Clyde, M. A., Palmieri Weber, R., Iversen, E. S., Poole, E. M., Doherty, J. A., Goodman, M. T., et al. (2016). Risk prediction for epithelial ovarian cancer in 11 United States-based case-control studies: Incorporation of epidemiologic risk factors and 17 confirmed genetic loci. *American Journal of Epidemiology, 184*(8), 555–569.

Daniels, K., Daugherty, J., & Jones, J. (2014). *Current contraceptive status among women aged 15–44: United States* (pp. 2011–2013). Hyattsville, MD: NCHS data brief. https://doi.org/10.1016/j.contraception.2013.08.004.

Doherty, J. A., Peres, L. C., Wang, C., Way, G. P., Greene, C. S., & Schildkraut, J. M. (2017). Challenges and opportunities in studying the epidemiology of ovarian cancer subtypes. *Current Epidemiology Reports, 4*(3), 211–220. https://doi.org/10.1007/s40471-017-0115-y.

Farley, J. H., Tian, C., Rose, G. S., Brown, C. L., Birrer, M., & Maxwell, G. L. (2009). Race does not impact outcome for advanced ovarian cancer patients treated with cisplatin/paclitaxel: An analysis of gynecologic oncology group trials. *Cancer, 115*(18), 4210–4217. https://doi.org/10.1002/cncr.24482.

Fuh, K. C., Shin, J. Y., Kapp, D. S., Brooks, R. A., Ueda, S., Urban, R. R., et al. (2015). Survival differences of Asian and Caucasian epithelial ovarian cancer patients in the United States. *Gynecologic Oncology, 136*(3), 491–497.

Gnagy, S., Ming, E. E., Devesa, S., Hartge, P., & Whittemore, A. S. (2000). Declining ovarian cancer rates in U.S. women in relation to parity and oral contraceptive use. *Epidemiology, 11*(2), 102–105.

Hall, M. J., Reid, J. E., Burbidge, L. A., Pruss, D., Deffenbaugh, A. M., Frye, C., et al. (2009). BRCA1 and BRCA2 mutations in women of different ethnicities undergoing testing for hereditary breast-ovarian cancer. *Cancer, 115*(10), 2222–2233. https://doi.org/10.1002/cncr.24200.

Hess, V., A'Hern, R., Nasiri, N., King, D. M., Blake, P. R., Barton, D. P. J., et al. (2004). Mucinous epithelial ovarian cancer: A separate entity requiring specific treatment. *Journal of Clinical Oncology, 22*(6), 1040–1044.

Howlader, N., Noone, A., Krapcho, M., Miller, D., Bishop, K., Kosary, C., et al. (2017). *SEER cancer statistics review, 1975–2014*. Bethesda, MD: National Cancer Institute. based on November 2016 SEER data submission, posted to the SEER web site, April 2017, Retrieved June 22, 2017, from http://seer.cancer.gov/csr/1975_2014/.

Ibeanu, O. A., & Díaz-Montes, T. P. (2013). Outcomes in ovarian cancer among hispanic women living in the United States: A population-based analysis. *Pathology Research International, 2013*, 10–15.

John, E. M., Whittemore, A. S., Harris, R., & Itnyre, J. (1993). Characteristics relating to ovarian cancer risk: Collaborative analysis of seven U.S. case-control studies. Epithelial ovarian cancer in black women. Collaborative Ovarian Cancer Group. *Journal of the National Cancer Institute, 85*(2), 142–147.

Jordan, S. J., Nagle, C. M., Coory, M. D., Maresco, D., Protani, M. M., Pandeya, N. A., et al. (2013). Has the association between hysterectomy and ovarian cancer changed over time? A systematic review and meta-analysis. *European Journal of Cancer (Oxford, England: 1990), 49*(17), 3638–3647.

Konecny, G. E., Wang, C., Hamidi, H., Winterhoff, B., Kalli, K. R., Dering, J., et al. (2014). Prognostic and therapeutic relevance of molecular subtypes in high-grade serous ovarian cancer. *Journal of the National Cancer Institute*, *106*(10), dju249. https://doi.org/10.1093/jnci/dju249.

Kurian, A. W. (2010). BRCA1 and BRCA2 mutations across race and ethnicity: Distribution and clinical implications. *Current Opinion in Obstetrics and Gynecology*, *22*(1), 72–78. https://doi.org/10.1097/GCO.0b013e328332dca3.

Kurman, R. J., Carcangiu, M. L., Herrington, C. S., & Young, R. H. (2014). *WHO classification of tumours of female reproductive organs* (4th ed.). Lyon: IARC.

Martin, J. A., Hamilton, B. E., Osterman, M. J. K., Driscoll, A. K., & Mathews, T. J. (2017). National vital statistics reports births: Final data for 2015. In *Vol. 66. National vital statistics reports*. Hyattsville, MD.

Moorman, P. G., Palmieri, R. T., Akushevich, L., Berchuck, A., & Schildkraut, J. M. (2009). Ovarian cancer risk factors in African-American and white women. *American Journal of Epidemiology*, *170*(5), 598–606.

Ness, R. B., Grisso, J. A., Klapper, J., & Vergona, R. (2000). Racial differences in ovarian cancer risk. *Journal of the National Medical Association*, *92*(4), 176–182.

Park, H. K., Ruterbusch, J. J., & Cote, M. L. (2017). Recent trends in ovarian Cancer incidence and relative survival in the United States by race/ethnicity and histologic subtypes. *Cancer Epidemiology, Biomarkers & Prevention*, *26*(10), 1511–1518.

Pectasides, D., Pectasides, E., Psyrri, A., & Economopoulos, T. (2006). Treatment issues in clear cell carcinoma of the ovary: A different entity? *The Oncologist*, *11*, 1089–1094.

Peres, L. C., Alberg, A. J., Bandera, E. V., Barnholtz-Sloan, J., Bondy, M. L., Cote, M. L., et al. (2017). Premenopausal hysterectomy and risk of ovarian cancer in African American women. *American Journal of Epidemiology*, *186*, 46–53. https://doi.org/10.1093/aje/kwx055.

Peres, L. C., Cushing-Haugen, K. L., Köbel, M., Harris, H. R., Berchuck, A., Rossing, M. A., et al. (2019). Invasive epithelial ovarian cancer survival by histotype and disease stage. *Journal of the National Cancer Institute*, *111*, 60–68. https://doi.org/10.1093/jnci/djy071.

Peres, L. C., Risch, H., Terry, K. L., Webb, P. M., Goodman, M. T., Wu, A. H., et al. (2018). Racial/ethnic differences in the epidemiology of ovarian cancer: A pooled analysis of 12 case-control studies. *International Journal of Epidemiology*, *47*(2), 460–472. https://doi.org/10.1093/ije/dyx252.

Peterson, C. E., Rauscher, G. H., Johnson, T. P., Kirschner, C. V., Barrett, R. E., Kim, S., et al. (2014). The association between neighborhood socioeconomic status and ovarian cancer tumor characteristics. *Cancer Causes & Control: CCC*, *25*(5), 633–637. https://doi.org/10.1007/s10552-014-0357-7.

Phelan, C. M., Kuchenbaecker, K. B., Tyrer, J. P., Kar, S. P., Lawrenson, K., Winham, S. J., et al. (2017). Identification of 12 new susceptibility loci for different histotypes of epithelial ovarian cancer. *Nature Genetics*, *49*(5), 680–691. https://doi.org/10.1038/ng.3826.

Prat, A., Pineda, E., Adamo, B., Galván, P., Fernández, A., Gaba, L., et al. (2015). Clinical implications of the intrinsic molecular subtypes of breast cancer. *Breast (Edinburgh, Scotland)*, *24*(Suppl. 2), S26–S35. https://doi.org/10.1016/j.breast.2015.07.008.

Reis-Filho, J. S., & Pusztai, L. (2011). Gene expression profiling in breast cancer: Classification, prognostication, and prediction. *Lancet (London, England)*, *378*(9805), 1812–1823. https://doi.org/10.1016/S0140-6736(11)61539-0.

Schildkraut, J. M., Alberg, A. J., Bandera, E. V., Barnholtz-Sloan, J., Bondy, M., Cote, M. L., et al. (2014). A multi-center population-based case-control study of ovarian cancer in African-American women: The African American Cancer Epidemiology Study (AACES). *BMC Cancer*, *14*(688), 688. https://doi.org/10.1186/1471-2407-14-688.

Schildkraut, J. M., Iversen, E. S., Akushevich, L., Whitaker, R., Bentley, R. C., Berchuck, A., et al. (2013). Molecular signatures of epithelial ovarian cancer: Analysis of associations with tumor characteristics and epidemiologic risk factors. *Cancer Epidemiology, Biomarkers & Prevention: A Publication of the American Association for Cancer Research, cosponsored by the American Society of Preventive Oncology, 22*(10), 1709–1721. https://doi.org/10.1158/1055-9965.EPI-13-0192.

Shida, D., Takabe, K., Kapitonov, D., Milstein, S., & Spiegel, S. (2008). Targeting SPHK1 as a new strategy against cancer. *Current Drug Targets, 9*(8), 662–673 (12).

Siegel, R. L., Miller, K. D., & Jemal, A. (2017). Cancer statistics, 2017. *CA: a Cancer Journal for Clinicians, 67*(1), 7–30.

Song, H., Ramus, S. J., Tyrer, J., Bolton, K. L., Gentry-Maharaj, A., Wozniak, E., et al. (2009). A genome-wide association study identifies a new ovarian cancer susceptibility locus on 9p22.2. *Nature Genetics, 41*(9), 996–1000.

Sopik, V., Iqbal, J., Rosen, B., & Narod, S. A. (2015). Why have ovarian cancer mortality rates declined? Part II. Case-fatality. *Gynecologic Oncology, 138*(3), 750–756. https://doi.org/10.1016/j.ygyno.2015.06.016.

Takano, M., Tsuda, H., & Sugiyama, T. (2012). Clear cell carcinoma of the ovary: Is there a role of histology-specific treatment? *Journal of Experimental & Clinical Cancer Research, 31*(1), 53. https://doi.org/10.1186/1756-9966-31-53.

Terplan, M., Schluterman, N., McNamara, E. J., Tracy, J. K., & Temkin, S. M. (2012). Have racial disparities in ovarian cancer increased over time? An analysis of SEER data. *Gynecologic Oncology, 125*(1), 19–24.

The Cancer Genome Atlas Research Network. (2011). Integrated genomic analyses of ovarian carcinoma. *Nature, 474*(7353), 609–615.

Tothill, R. W., Tinker, A. V., George, J., Brown, R., Fox, S. B., Lade, S., et al. (2008). Novel molecular subtypes of serous and endometrioid ovarian cancer linked to clinical outcome. *Clinical Cancer Research, 14*(16), 5198–5208.

Warren, J., Cronin, K. A., Stevens, J., Howlader, N., Trimble, E. L., & Harlan, L. C. (2017). Racial disparities in receipt of guideline care and cancer deaths for women with ovarian cancer. *Journal of Clinical Oncology, 35*(15_suppl), e18081. https://doi.org/10.1200/JCO.2017.35.15_suppl.e18081.

Webb, P. M., Green, A. C., & Jordan, S. J. (2017). Trends in hormone use and ovarian cancer incidence in US white and Australian women: Implications for the future. *Cancer Causes & Control, 28*(5), 365–370. https://doi.org/10.1007/s10552-017-0868-0.

Wentzensen, N., Poole, E. M., Trabert, B., White, E., Arslan, A. A., Patel, A. V., et al. (2016). Ovarian cancer risk factors by histologic subtype: An analysis from the Ovarian Cancer Cohort Consortium. *Journal of Clinical Oncology, 34*(24), 2888–2898.

Wu, A. H., Pearce, C. L., Tseng, C. C., & Pike, M. C. (2015). African Americans and hispanics remain at lower risk of ovarian cancer than non-hispanic whites after considering nongenetic risk factors and oophorectomy rates. *Cancer Epidemiology, Biomarkers and Prevention, 24*(7), 1094–1100.

Yang, H. P., Anderson, W. F., Rosenberg, P. S., Trabert, B., Gierach, G. L., Wentzensen, N., et al. (2013). Ovarian cancer incidence trends in relation to changing patterns of menopausal hormone therapy use in the United States. *Journal of Clinical Oncology, 31*(17), 2146–2151. https://doi.org/10.1200/JCO.2012.45.5758.

Zeng, C., Wen, W., Morgans, A. K., Pao, W., Shu, X.-O., & Zheng, W. (2015). Disparities by race, age, and sex in the improvement of survival for major cancers. *JAMA Oncology, 1*(1), 88–96. https://doi.org/10.1001/jamaoncol.2014.161.

CHAPTER TWO

Age-related disparities in older women with breast cancer

Nusayba A. Bagegni, Lindsay L. Peterson*
Washington University in St. Louis School of Medicine, St. Louis, MO, United States
*Corresponding author: e-mail address: llpeterson@wustl.edu

Contents

Abstract

Improvements in breast cancer (BC) mortality rates have not been seen in the older adult community, and the fact that older adults are more likely to die from their cancer than younger women establishes a major health disparity. Studies have identified that despite typically presenting with more favorable histology, older women present with more advanced disease, which may be related in part to delayed diagnosis. This is supported by examination of screening practices in older adults. Older women have a worse prognosis than younger women in both early stage disease, and more advanced and metastatic disease. Focus on the treatment of older adults has often concentrated on avoiding overtreatment, but in fact undertreatment may be one reason for the age-related differences in outcomes, and treatments need to be individualized for every older adult, and take into account patient preferences and functional status and not chronologic age alone. Given the aging population in the US, identifying methods to improve early diagnosis in this population and identify additional factors will be important to reducing this age-related disparity.

Advances in Cancer Research, Volume 146
ISSN 0065-230X
https://doi.org/10.1016/bs.acr.2020.01.003

1. Introduction

Breast cancer (BC) is the most commonly diagnosed malignancy in women and one of the leading causes of cancer mortality worldwide (Wildiers et al., 2007). Overall, approximately one in eight women will be diagnosed with invasive BC within her lifetime (http://seer.cancer.gov/statfacts/html/breast.html, accessed August 2017). An estimated 252,710 new cases of invasive BC will be diagnosed in the United States (U.S.) in 2017, accounting for over 16 billion dollars in medical costs (http://seer.cancer.gov/statfacts/html/breast.html;Mariotto, Yabroff, Shao, Feuer, & Brown, 2011). Moreover, based on estimates from the Surveillance, Epidemiology, and End Results (SEER) Program database, the largest projected increase in cancer care costs by 2020 was in the continuing phase of care for female BC and prostate cancer, with female BC accounting for 32% of the cost (Mariotto et al., 2011).

Age remains the major risk factor for BC development in women (Muss et al., 2009). Nearly one third of all patients diagnosed with BC are 70 years of age or older (Ring et al., 2011). There is a 6.8% lifetime risk of developing BC in women of this age group compared to 3.5% in women ages 60–69 and 1.9% in women <49 years of age (https://cancerstatisticscenter.cancer.org/#/cancer-site/Breast (accessed August 2017), https://www.cancer.org/cancer/breast-cancer/about/how-common-is-breast-cancer.html, accessed August 2017). Improvements in screening and advances in curative intent therapies have improved BC mortality rates over time. However, survival among women diagnosed in older age has not substantially improved (Berry et al., 2005; Smith et al., 2011). Older adults with BC are more likely to die of their disease than younger women (Singh, Hellman, & Heimann, 2004) and nearly half of BC deaths occur in women age 70 or older (Orucevic et al., 2015). This age-related disparity in prognosis has largely been attributed to advanced disease at presentation related to delays in time to diagnosis (Turner, Zafarana, Becheri, Mottino, & Biganzoli, 2013). Older women are more likely to present with larger tumor size and greater lymph node involvement (Botteri, Bagnardi, Goldhirsch, Viale, & Rotmensz, 2010; Gennari et al., 2004; Wildiers et al., 2007), which supports this theory.

It is estimated that about 20% of the U.S. population will be older than 65 years by 2030 (Barginear et al., 2014). The average 65- and 75-year old patient has a 20- and 12-year estimated life expectancy, respectively (Barginear et al., 2014). With this shift in demographics and increasing life expectancy, the proportion of older women diagnosed with BC is expected to substantially rise. In fact, approximately 70,000 women over the age of 70 are diagnosed with BC in the U.S. annually (Freedman & Partridge, 2017), and the largest increase in the total number of cancer survivors in 2020 is projected for those aged 65 years and older (Mariotto et al., 2011). A comprehensive understanding of relevant survivorship issues and management of any long-term toxicity from BC treatment in this aging cohort therefore becomes imperative. The oncology community must therefore be aware of the age-related disparities in BC and strive to improve prognosis for older women diagnosed with this disease. The purpose of this review is to highlight the age-related disparity in BC prognosis, summarize factors that contribute to this disparity, and discuss strategies to improve prognosis in older adults with BC.

2. Prognosis in older breast cancer survivors

BC incidence dramatically increases with age and will rise with the expanding population of older adults and increasing life expectancy. As older adults make up a larger proportion of new BC diagnoses, it is important to understand the age-related disparities in BC. Extremes of age (<40 and >70) are associated with worse prognosis in women with BC compared to middle-aged women (aged 40–69), with advanced age being a significant independent predictor of poor BC prognosis (Chen, Zhou, Tian, Meng, & He, 2016). Although the majority of older women with early-stage disease die from alternative causes, the risk of BC-related death increases with age, most significantly after the age of 80 (VanderWalde & Hurria, 2012). Women over the age of 85 have a mortality risk 13 times that of younger women (VanderWalde & Hurria, 2012). Overall, approximately 57% of deaths resulting from BC occur in patients over the age of 65. The probability of BC-related death for women ages <49, 60–69 or 70 and older is 0.2, 0.6 and 2%, respectively (https://cancerstatisticscenter.cancer.org/#/cancer-site/Breast). In a recent study comparing outcomes in non-metastatic BC patients with triple negative disease by age, advanced age was a predictor of worse outcomes, with women aged ≥70 having worse cancer-specific and overall survival compared to younger women with triple negative BC (aged <70 years) (Zhu, Perez, Hong, Li, & Xu, 2015). Furthermore, older women presenting with locally advanced or metastatic BC are more likely to have worse outcomes than younger women with advanced disease (Freedman & Partridge, 2017). Approximately 48% of patients over the age of 65 present with metastatic disease at the time of diagnosis (de novo metastatic disease) compared to an estimated 26% in all patients (Mariotto, Etzioni, Hurlbert, Penberthy, & Mayer, 2017; Tesarova, 2016). Among women with de novo metastatic disease, older women have worse survival rates than younger women (Mariotto et al., 2017). To address these disparities in prognosis, the approach to treating the older adult with BC should focus on tumor biology, as well as the functional status and comorbidities of the patient, rather than chronologic age alone.

3. Age-related differences in tumor biology

Epidemiologic studies of BC suggest important age-related differences in the molecular and biological phenotype of BC with advancing age. In general, BC in older patients is a more indolent disease, with favorable features including lower histologic grade, lower proliferative indices, less peritumoral vascular invasion and more well-differentiated tumors as compared to younger patients (Le Saux et al., 2015; Wildiers et al., 2007). The majority of tumors express estrogen receptor (ER) and/or progesterone receptor (PR), and lack human epidermal growth factor receptor 2 (HER-2) overexpression (Blair, Robles, Weiss, Ward, & Unkart, 2016). In fact, over 80% of tumors diagnosed in women ages 80–84 are hormone receptor positive (Kuijer & King, 2017). This is slightly higher than the rates of hormone receptor positive BC in the overall population. Although ductal histology remains the most prevalent histology in older patients, lobular carcinomas and carcinomas that tend to have more favorable prognosis, such as mucinous and papillary carcinomas, are slightly overrepresented in the older population (Wildiers et al., 2007). Tumors generally exhibit normal p53 expression (Tesarova, 2016) (multiple studies illustrate the absence of mutated p53 is predictive of longer disease-free survival and overall survival following initial therapy for BC) (Yang, Du, Kwan, Liang, & Zhang, 2013). Studies using the PAM50 gene assay to report molecular subtypes show rising incidence of luminal A tumors and decreasing incidence of luminal B, HER2 enriched and basal-like molecular subtypes with increasing age. Nevertheless, one study demonstrated that about 19% of older women have luminal B tumors, which are typically associated with more aggressive biology (VanderWalde & Hurria, 2012). Similar to younger women, HER2 positive BC and triple negative BC serve as poor prognostic factors affecting BC specific survival in older adults compared to luminal type tumors (Königsberg et al., 2016). Contrary to the more favorable features common to BC in older adults, tumor size increases with age, particularly after age 80 years (Schonberg et al., 2010). Although larger tumor size at diagnosis is a poor prognostic feature, this is likely a reflection of screening practices in older adults as opposed to a sign of more aggressive disease in general (Blair et al., 2016). This is supported by the fact that ductal carcinoma in situ (DCIS) represents a lower proportion of cancers among those aged 65 years or older compared to women ages 40–64 years (Henderson, O'Meara, Braithwaite, & Onega, 2015).

4. Breast cancer screening in older women

Advanced age is the most important risk factor for development of female BC. Breast cancers generally have a latency period; thus, screening mammography has contributed to early detection when it has a higher probability of cure. Beginning in the 1980s, there has been a substantial increase in the use of screening mammography among women aged 40 and older (Bleyer & Welch, 2012). Based on randomized screening trials, screening mammography is estimated to result in a 15–25% relative risk reduction in BC mortality in women aged 50–69 years (Braithwaite, Demb, & Henderson, 2016; Løberg, Lousdal, Bretthauer, & Kalager, 2015). Observational studies have suggested that screening mammography on a biannual basis for 10 years reduces mortality by 0.2% in women above the age of 70 (Tesarova, 2016). Although data among retrospective population-based studies is conflicting, some studies suggest the use of screening mammography improved BC or all-cause mortality in older women except in the setting of severe comorbid conditions (Walter & Schonberg, 2014). A recent analysis of various models examining the risks and benefits of different screening recommendations (annual screening at ages 40–84 years; screening annually at ages 45–54 years, then biennially at ages 55–79 years; and biennial screening at ages 50–74 years) found the greatest mortality reduction in annual screening at ages 40–84 (Arleo, Hendrick, Helvie, & Sickles, 2017). However, there are several reasons why healthcare providers may be less likely to refer older adults for screening mammograms.

Data surrounding the specific benefits and harms of screening mammography for older women are lacking. Prospective randomized trials of screening mammography have not included women over the age of 75, resulting in vague or uncertain guidelines in this population (Walter & Schonberg, 2014). The limited availability of evidence-based data serves as the basis for the 2009 U.S. Preventive Services Task Force

(USPSTF) updated guidelines concluding that the evidence is insufficient to fully assess the benefits and harms of screening mammography in women over age 75 (USPSTF, 2009). Such guidelines, along with the lack of prospective data, likely explain the exceedingly variable initial and subsequent BC screening patterns described in older women. In fact, a retrospective study demonstrated declining rate of subsequent screening among female Medicare beneficiaries age 65 and older screened between 2004 and 2009, with older women being less likely to return for subsequent screening (Jiang, Hughes, Appleton, McGinty, & Duszak, 2015).

There are potential consequences of screening. These include patient anxiety and psychological distress as well as the possible need for additional diagnostic imaging studies, invasive biopsies and higher use of health care services related to false-positive results. In a study of breast and colorectal cancer screening in women aged 50–80, psychological distress associated with screening was one of the strongest negative predictive barriers to screening (O'Donnell et al., 2010). The sensitivity and specificity of mammography improves with age, highest in women over the age of 80 (Schonberg et al., 2010). Thus, women with good functional status and without significant comorbidities may benefit from ongoing screening (Braithwaite et al., 2016; Nelson et al., 2009; Oeffinger et al., 2015). However, although the highest rate of false-positive mammography results have been reported among women age 40–49 and decreases with age, false-positive results remain common in all age groups (Nelson, O'Meara, Kerlikowske, Balch, & Miglioretti, 2016). Also, the phenomena of overdiagnosis, referring to the diagnosis of a screen-detected cancer that is unlikely to become symptomatic or affect a patient's life expectancy, must also be considered given mammography may detect in situ carcinomas as well as indolent invasive tumors in patients with other life-limiting diagnoses. The magnitude of overdiagnosis (which can lead to unnecessary treatment) is difficult to precisely estimate and can only be inferred, with published literature suggesting rates ranging from less than 5% to >50% (Bleyer & Welch, 2012; Løberg et al., 2015; Nelson et al., 2009; Oeffinger et al., 2015). It is estimated that approximately 30% of screening mammography-detected cancers would not have otherwise impacted life expectancy in women over the age of 75 (Schonberg, 2016). Thus, given the fact that the majority of older patients will die of non-BC-related comorbid causes, overdiagnosis and overtreatment could carry a higher burden of harm than for younger patients (Freedman, Keating, Partridge, et al., 2017).

Collectively, such observations are addressed by the American Cancer Society recommendation of taking an individualized approach in determining screening in older patients with average BC risk, recommending that women continue screening mammography if their predicted life expectancy is 10 years or longer (Oeffinger et al., 2015). Several other guidelines also support an individualized approach to screening mammography in older women (Walter & Schonberg, 2014). Although shorter life expectancy is associated with increasing age, suggesting a reduction in absolute benefit from screening, epidemiologic studies demonstrate an improvement in life expectancy over time with an increasingly aging population. In the U.S., the average life expectancy for a woman aged 65, 75 and 85 years is 86.7, 88.6 and 92.3 years, respectively (https://www.ssa.gov/OACT/population/longevity.html, accessed September 2017). Uniform guidelines based on chronologic age alone are likely to contribute to overdiagnosis in some cases, and undertreatment in others. Thus, screening decisions based on age alone has the possibility of harming some older adults who may actually have life expectancies of 5, 10, or even 15 years.

Surprisingly, older BC survivors appear less likely to receive routine follow up breast imaging as well. In a study of women aged ≥65 who underwent primary surgical therapy for stage I–II BC (1992–1999), those who were older were less likely to undergo annual post-diagnosis surveillance mammography (Keating, Landrum, Guadagnoli, Winer, & Ayanian, 2006). This practice was again observed in a more recent population-based analysis of older BC survivors, where the use of surveillance mammography decreased with advancing age and declining life expectancy (Freedman, Keating, Partridge, et al., 2017). In fact, one study found 14.1% of older BC survivors did not undergo surveillance mammography despite having a life expectancy of >10 years (Freedman, Keating, Pace, et al., 2017).

It is critical to balance the potential benefits and harms of screening in the context of advancing screening technologies. New strategies are needed to help guide screening practices that most accurately tailor to an individual patient's likelihood to benefit from continued screening based on burden of comorbid conditions, functional status, estimated life expectancy and potential risks of treatment. Prospective studies incorporating validated comorbidity indices to pursue life expectancy based screening will help shed light on this important issue (Braithwaite et al., 2016). Various web-based calculators are available to calculate life expectancy (https://eprognosis.ucsf.edu/ (accessed September 2017); https://www.ssa.gov/OACT/population/longevity.html (accessed September 2017)).

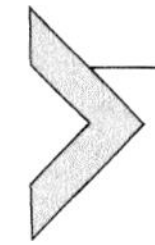

5. Breast cancer treatment in older women

5.1 Treatment considerations

Despite studies demonstrating that older women with BC have similar survival benefits from curative intent therapies as younger women (Singh et al., 2004), undertreatment of older adults is likely contributing to the age-related disparity in BC prognosis. Numerous studies have described patterns of less aggressive treatment in older BC patients, with less likelihood of receiving locoregional and systemic treatments according to consensus treatment guidelines (Bouchardy et al., 2003; EBCTCG et al., 2012; Yancik et al., 2001). One explanation for this is the high prevalence of comorbid conditions in the elderly. Older adults with cancer tend to have more coexisting chronic conditions compared to age-matched controls without cancer, with more than half having at least one condition that may impact cancer treatment (Williams et al., 2016). This is true in BC as well, where the prevalence of pre-existing comorbid conditions increases with age, impacting nearly 70% of BC patients older than age 80 (Ogle, Swanson, Woods, & Azzouz, 2000; Yancik, 1997).

The most frequent concomitant conditions in older adults with cancer include prior cancer, chronic obstructive pulmonary disease, diabetes mellitus, hypertension and heart disease (Janssen-Heijnen et al., 2005). It is well established that BC patients with comorbidities have a poorer prognosis than patients without comorbidities, and comorbidities can be as crucial as cancer stage in predicting BC survival (Patnaik, Byers, Diguiseppi, Denberg, & Dabelea, 2011). Breast cancer patients with three or more comorbid illnesses, notably diabetes, diseases of the lung, liver, or cardiovascular system or concomitant malignancy, have a 20-fold higher mortality rate from causes other than BC (Balducci & Yates, 2000). In the Dutch OMEGA study, increased number of geriatric comorbid conditions as assessed by pretreatment comprehensive geriatric assessment strongly correlated with increased grade 3–4 chemotherapy-related toxicity in older women with metastatic BC treated with pegylated liposomal doxorubicin or capecitabine (Hamaker et al., 2014). Surveillance, Epidemiology, and End Results Program (SEER)-based studies have also shown that older women aged $\geq$70 years with non-metastatic BC are more likely to die from a serious comorbid illness (Muss, 2010). In a large SEER population-based study assessing impact of comorbid conditions on mortality in BC patients, overall survival rates in a univariate analysis were highest for patients of younger age, earlier tumor stage, lower tumor grade, ER-positive status and having no comorbidities (Patnaik et al., 2011). The presence of comorbidities was substantially associated with decreased survival. In fact, those diagnosed with early stage BC and with concomitant comorbidities had similar or worse survival outcomes to patients diagnosed with later-stage BC without identified comorbidities.

A better understanding of the impact of specific high risk comorbid conditions on cancer-specific survival is critical in guiding treatment decisions in older patients with early stage BC. Additionally, assessing frailty, a broad term characterized by increased vulnerability to declining physiologic and functional status, has become an increasingly recognized syndrome in the field of geriatric oncology and should be considered early on in the treatment course (Huisingh-Scheetz & Walston, 2017). It is predictive of surgical outcomes and chemotherapy intolerance. Frailty takes into account functional or cognitive impairment, poor nutritional status, mood disorders and advanced age (Huisingh-Scheetz & Walston, 2017). Other geriatric syndromes include falls and polypharmacy (Kirkhus et al., 2017). Polypharmacy has been associated with increased chemotherapy toxicity (Hamaker et al., 2014) as it increases the risk of drug interactions and can impact organ function, potentially impacting a patient's ability to tolerate treatment. This highlights the importance of proper management of concomitant comorbid conditions and recognition of frailty not only to optimize physical function and quality of life but also to improve prognosis by way of maximizing the chance of delivering curative intent therapies. A geriatric assessment, endorsed by the Society of International Geriatric Oncology expert panel, can objectively assess key domains associated with aging, identify deficits that may not otherwise be recognized by a routine oncologic evaluation of routine assessment of performance status, and can thereby be addressed early on in the course of cancer management (Mohile et al., 2016; Shachar, Hurria, & Muss, 2016). When time allows, referral to a geriatrician or appropriate sub-specialty provider to address frailty and optimize the management of comorbid conditions prior to and/or during cancer therapy is ideal. A variety of valuable assessment tools have become available to formally screen for and delineate high risk features in older adults (Huisingh-Scheetz & Walston, 2017), and assess the need for further evaluation with a comprehensive geriatric assessment.

Additional factors influencing patterns of BC care in older women are less well-characterized but include patient and/or family preference, undervalued life expectancy, health literacy, functional and cognitive health, socioeconomic factors, access to health care, and physician's reluctance or perception of treatment utility and safety (Mandelblatt et al., 2000; Schonberg, Silliman, McCarthy, & Marcantonio, 2014; Tesarova, 2013). There remains limited data defining the full impact, interplay and complexities of such unique factors on treatment decision-making in older adults with BC, undermining a comprehensive understanding of the clinical problem.

5.2 Patterns of locoregional treatments

Older women receive less aggressive treatment and experience higher mortality from early-stage BC than younger women (Schonberg et al., 2010). This is particularly true for women over the age of 80, who are less likely to receive surgery or undergo breast-conserving surgery alone (without radiation) as compared to those age 67–79 years. In fact, in one series, nearly 30% of women age ≥80 years did not receive standard locoregional therapy for stage I–II BC as compared to only 6% of younger women (Schonberg et al., 2010). Undertreatment and deviation from standard treatment guidelines is a well-described reality in older BC patients, which may be attributed to patient and/or physician concern of effect of comorbidities on treatment-related toxicity, limited life expectancy, or perceived good prognosis of more favorable breast cancer biology in older BC patients (Markopoulos & van de Water, 2012).

5.2.1 Surgery

Large population-based studies have shown that BC patients younger than 80 were more likely to undergo surgery than those older than 80 years (Janssen-Heijnen et al., 2005). A large prospective study of older women (aged >70 years) with operable Stage I–III BC found that women aged ≥85 years had lower likelihood of receiving BC surgery even after adjusting for patient choice and functional status (OR 0.18, 95% CI 0.07–0.44) (Tesarova, 2016). In a large cohort, the proportion of breast-conserving surgery decreased from 54% of those <60 years of age to 29% of those ≥80 years (Janssen-Heijnen et al., 2005). Patients with more comorbid conditions were less likely to undergo surgery (as well as radiotherapy). Importantly, patients 70 years and older did not have higher, but rather similar, complications rates as their younger counterparts (ages 40–69) when undergoing surgery, including minor infection or lymphedema post axillary dissection. In another large study, increasing age and increased Charlson comorbidity score (particularly age ≥80 years and comorbid score ≥1) were two variables predictive for not receiving combined locoregional treatment ($P < 0.01$), although increased age did not correlate with increased Charlson comorbidity index in this study ($P = 0.48$) (Hurria et al., 2003).

One study suggests that the risk of lymph node involvement decreases with age up to 70 years, but increases after the age of 70 years, primarily associated with smaller tumors (Wildiers et al., 2009). Among women potentially eligible for ACOSOG Z0011 (a randomized trial of axillary node dissection in women with clinical T1–2, N0, M0 BC with a positive lymph node), patients aged ≥70 years were more likely to undergo sentinel lymph node biopsy than axillary lymph node dissection (27.0% vs 20.1%; $P < 0.001$) (Ong et al., 2017). A prior randomized study of primary surgery vs. addition of axillary clearance (sentinel and axillary node dissection) in women with clinically node-negative BC aged ≥60 (1993–2002) plus adjuvant tamoxifen for ER-positive disease demonstrated similar 6-year disease-free and overall survival ($P = 0.77$) with superior early quality of life without axillary node dissection (International Breast Cancer Study Group et al., 2006).

Although reconstruction rates overall are on the rise, older women are less likely to undergo breast reconstruction post-mastectomy, possibly in part due to concern for increased surgical risk (Butz et al., 2015; Oh et al., 2016), or patient reluctance to undergo additional surgical procedures (Maly et al., 2009). One large study showed that only 10.7% of older women (≥65 years) received post-mastectomy reconstruction as compared to 39.5% of younger women (<65 years) (Butz et al., 2015). However, multiple studies demonstrate that advancing age is not a significant predictor of peri-operative complications following mastectomy and breast reconstruction (Santosa et al., 2016). Specifically, the risk of longer hospital stay, complication rate and re-operation rate were no different in women ≥65 than younger women (Butz et al., 2015). Age, however, was found to be an independent risk factor for venous thromboembolism after autologous reconstruction in one series (OR, 3.67; $P = 0.02$), which otherwise found no difference in surgical morbidity following implant-based reconstruction (Butz et al., 2015). Comorbid conditions, including obesity, renal insufficiency and impaired patient functional status, have been shown to increase the risk of venous thromboembolism following breast reconstruction above that of mastectomy, suggesting the importance of consideration of patient-specific factors in risk stratification (Fischer et al., 2014). Interestingly, older women report higher physical and psychosocial satisfaction than younger women with autologous reconstruction post-mastectomy

(Santosa et al., 2016). Additionally, patient reported quality of life outcome measures are similar regardless of age (Oh et al., 2016). An Australian study of patient reported outcomes following breast reconstruction in women aged ≥60 years requiring mastectomy for early stage BC found that the number of comorbidities was associated with the reason "My surgeon is against breast reconstruction" ($P=0.018$) in those who did not receive breast reconstruction (Oh et al., 2018). Likewise, age was associated with patient reported reason "I am too old to have breast reconstruction" for not receiving breast reconstruction ($P=0.001$). This highlights the need for clinicians to address perceived misconceptions about increased surgical risk in older women who are otherwise clinically fit for post-mastectomy breast reconstruction. Referral to a plastic surgeon for individualized decisions regarding reconstruction may be warranted. Additionally, guidelines for older adults undergoing surgery have been published by both the American Geriatrics Society and the American College of Surgeons and may be helpful in surgical planning of older adults with BC (Chow et al., 2012).

5.2.2 Radiation therapy

Post-operative irradiation remains the standard of care following breast-conserving surgery to treat early stage BC. In a large meta-analysis, breast radiotherapy post breast-conserving surgery resulted in a significant reduction in the 15-year risk of BC death (absolute reduction of 3.8%) (EBCTCG et al., 2011). In women aged 70 and older, post lumpectomy radiation resulted in lower rates of local ipsilateral recurrence but not overall survival (Hughes et al., 2013, 2004). However, in a recent retrospective SEER population-based study of women aged 70 or older who underwent breast-conserving surgery (1999–2003) for pT1a-b, N0 BC, adjuvant radiation therapy was associated with lower risk of BC death compared with patients who did not receive surgery (adjusted HR $=0.55$; $P<0.0001$) (Nagar et al., 2017). Additionally, a SEER population-based study (1998–2011) assessed long-term outcomes in women aged ≥70 years treated with and without adjuvant radiation therapy following breast conservation surgery for T1, ER-negative invasive ductal carcinoma (Daugherty et al., 2016). Patients treated with adjuvant radiation ($n=3685$) had a younger median age and improved 5-year overall survival of 81% compared to 61.7% for those not receiving adjuvant radiation ($n=1493$) ($P<0.0001$), as well as 5-year cancer-specific survival (93.1% vs 85%, $P<0.0001$). Likewise, another SEER-based retrospective study of the impact of adjuvant radiotherapy after breast-conserving surgery in older women (≥70 years) with T1–2, N0 ER-negative BC on incidence of future mastectomy and BC death found women receiving radiation were younger (aged <75) and more often had low Charlson-Deyo Comorbidity Index (41% vs 25%) (Eaton et al., 2016). In this study, adjuvant radiation was associated with a reduced incidence of future mastectomy (4.9% vs 8.3%) and BC death (10.8% and 24.1%, $P < 0.001$) (Eaton et al., 2016).

The multi-national randomized, prospective phase 3 PRIME II study sought to determine whether omission of adjuvant whole-breast radiotherapy (40–50 Gy in 15–25 fractions) affected ipsilateral breast tumor recurrence in older women (aged ≥65) with low risk hormone receptor positive, T1-T2 (≤3 cm), N0 non-metastatic BC treated with breast-conserving surgery, pathologic axillary staging and receiving adjuvant endocrine therapy (Kunkler et al., 2015). At median 5-year follow-up, the rate of ipsilateral breast tumor recurrence was higher in those not receiving post-operative radiotherapy (4.1%) than those assigned to radiotherapy (1.3%) ($P = 0.0002$); however, 5-year overall survival was identical in both groups (93.9%, $P = 0.34$) (Kunkler et al., 2015). The earlier CALGB 9343 trial randomized women ages ≥70 with very low risk hormone receptor positive BC (clinical stage I, T1 N0 M0, negative margins) to adjuvant tamoxifen with or without adjuvant whole-breast irradiation following breast-conserving surgery (Hughes et al., 2013, 2004). Once again, the rate of locoregional recurrence at 5 years varied between both groups (4% without adjuvant radiation vs 1% with radiotherapy, $P < 0.001$) but without statistically significant differences in rate of mastectomy for local recurrence, distant recurrence or overall survival, which remained following long-term follow-up (12.5 years) (Hughes et al., 2004). A variety of alternative studies showed similar findings. It is thus clear that for some women with low risk early stage BC, radiation therapy may be omitted without compromising overall survival. Additionally, although the rate of local recurrence is higher when radiation is omitted in these circumstances, the overall risk of local recurrence even without radiation is very low (4.1%) (Hughes et al., 2004). Although the literature suggests that advancing age is associated with more favorable tumor biology, data is limited in larger or higher-grade tumors and thus this finding cannot be extrapolated to all hormone receptor-positive tumors.

There is a paucity of evidence regarding the impact of comorbid conditions, and thereby life expectancy, on outcomes post-radiotherapy.

There are also unique considerations such as quality of life measures and inconveniences or stressors imposed by treatment schedules that must be considered when making individualized recommendations for or against radiation therapy. In one study, 80% of older patients were able to complete radiation therapy (Blair et al., 2016). Despite this, in one series, 49% of older patients refused adjuvant radiation (Vetter et al., 2013). It is not uncommon for patients to make assumptions that a treatment will not be beneficial or will not be tolerable. It is important to discuss radiation therapy recommendations in the context of the risks and benefits unique to every patient allowing for a shared decision-making process. If a patient declines radiation, the rationale for this should be explored, so as to ensure patients are adequately provided the information they need to make well-informed decisions.

5.3 Systemic therapy

Treatment for BC is multidisciplinary, based on patient and tumor characteristics. Systemic therapies including endocrine therapy, cytotoxic chemotherapy and anti-HER2 blockade are associated with improved outcomes in both the early stage and late stage (metastatic) settings. Despite studies suggesting older women generally derive similar benefits as their younger counterparts from the addition of systemic therapies for BC therapy, these treatments are often withheld due to a variety of competing factors.

5.3.1 Endocrine therapy

Endocrine therapy remains the mainstay of treatment in all hormone receptor positive BCs, particularly in older women where most tumors are hormone receptor positive. Options in both the adjuvant and metastatic setting include tamoxifen and the aromatase inhibitors (AIs). Tamoxifen has been shown to decrease the annual rate of BC recurrence by 51% regardless of patient age (VanderWalde & Hurria, 2012). Third generation AIs have provided significant improvement in outcomes in postmenopausal women (Nabholtz, 2008), with data from randomized controlled trials showing a consistent advantage of AIs to tamoxifen in this cohort of patients for both advanced and early stage BC (Schneider et al., 2011). Data from meta-analyses of adjuvant trials demonstrate that AIs have a more favorable risk/benefit profile than tamoxifen in postmenopausal women regardless of age (Chlebowski et al., 2015). The MA.17 trial of extended adjuvant endocrine therapy with letrozole following 5 years of tamoxifen found no difference in quality of life measures at 24 months in women

≥ 70 years of age (Muss et al., 2008). Large scale randomized adjuvant trials of AIs clearly identify an increased rate of skeletal disorders (Nabholtz, 2008). This is particularly important to consider in older women who have increased prevalence of osteopenia and osteoporosis compared to younger women. Close monitoring of bone mineral density with early initiation of therapies to augment bone health will be essential to prevent potentially detrimental adversities in this susceptible cohort (O'Connor et al., 2013).

Given the option now for prolonged duration of adjuvant endocrine-based treatment (up to 10 years), compliance with endocrine therapy remains a major concern as it can impact long-term outcomes. Studies assessing factors hindering full compliance have yielded mixed results. In one series, 13% of older patients chose to forgo adjuvant hormone therapy (Vetter et al., 2013). Although a large study of women with hormone-sensitive stage I–III BC assessing factors associated with suboptimal treatment adherence found that women younger than 40 years of age had the highest risk of discontinuation, both younger or advanced age as well as comorbidities was associated with earlier treatment cessation (Hershman et al., 2010). The Breast International Group 1–98 clinical trial found that reduced adherence was associated with older age in postmenopausal women randomized to tamoxifen or letrozole (Chirgwin et al., 2016). In the CALGB/Alliance 369901 study, frailty (assessed by adapted version of the frailty index developed by Searle et al., 2008) was associated with a higher risk of non-initiation of adjuvant endocrine therapy (OR 1.63, $P = 0.013$), and advanced age increased the risk of discontinuation ($P = 0.005$) (Sheppard et al., 2014). Another study investigating endocrine therapy compliance in BC patients diagnosed ≥ 80 years ($n = 79$) as compared to patients diagnosed at age 60–79 years ($n = 358$), the older cohort was more likely to decline recommended endocrine therapy (13.0% vs 4.5%, $P = 0.011$), and after initiating treatment, were less likely to complete the planned treatment duration (39.6% vs 71.3%, $P < 0.001$) (Güth et al., 2013). Interestingly, this study found that treatment was more often discontinued by the physician due to serious medical reasons other than BC (17.0% vs 4.7%, $P = 0.003$). Additionally, older women were more often treated by a general practitioner than women aged 60–79 (54.4% vs 23.9%, $P < 0.001$) (Güth et al., 2013). A recent pilot study to assess the use of a web-based application designed with and without weekly reminders for patients to report AI use and symptoms was found to improve medication adherence in postmenopausal women with early stage hormone receptor positive BC (Graetz et al., 2018).

Prior to discontinuing endocrine therapy, a thorough review of the risks and benefits should be discussed with patients. The benefit of endocrine therapy can be assessed by re-reviewing genomic assays completed at the time of diagnosis, or utilizing an on-line tool that estimates the risk reduction of endocrine therapy taking into account patient and tumor characteristics (http://www.lifemath.net/cancer/?cancer (accessed September 2017), www.predict.nhs.uk/predict.html, accessed September 2017, https://www.cts5-calculator.com/ (accessed February 2020)). Additionally, when therapy is discontinued for a bothersome side effect as opposed to a serious medical issue, re-evaluation after several weeks off the medication to ensure that the side effect was in fact caused by the medication is warranted. This will ensure that treatment is not discontinued for a side effect that is ultimately not attributable to the hormonal therapy.

5.3.1.1 Primary endocrine therapy

Older women tend to have hormone receptor positive tumors (ER-positive and/or PR-positive), suggesting a possible role for primary endocrine therapy, particularly in patients deemed unfit to undergo surgical resection due to extensive comorbid conditions or limited life expectancy. An analysis based on data from the National Cancer Database assessing trends in use of treatment modality (surgery and endocrine therapy) in over 95,000 women ≥80 years with invasive, ER-positive BC found that the rate of primary non-operative treatment doubled from 7% in 2004 to 14% in 2012 (Kantor et al., 2016). Multivariate logistic regression analysis identified more advanced clinical stage and older age as independent predictors of receiving primary non-surgical management.

A 2006 Cochrane review of seven randomized controlled trials comparing surgery to endocrine therapy in elderly (but fit to undergo surgery) women with operable BC showed that tamoxifen alone was inferior to surgery with respect to progression-free survival ($HR=0.55$, 95% $CI=0.39$–0.77, $P=0.0006$), but without significant difference in overall survival ($HR=0.98$, 95% $CI=0.74$–1.30, $P=0.9$) (Hind et al., 2006). A small study of primary hormone therapy for patients (median age 81) with ER-positive BC deemed medically unfit to undergo surgical resection or tolerate general anesthesia assessed treatment with primary endocrine therapy, including tamoxifen and/or aromatase inhibitors (Osborn et al., 2011). Of 82 patients (with 83 tumors) followed for a median of 24 months, 15% of patients had disease progression (2% within 6 months of diagnosis), 28% died and 17% ultimately underwent wide local excision under local anesthesia.

Another study in the Netherlands evaluated response and long-term outcomes of 184 women ages $\geq$75 years (median age 84 years) with operable BC receiving primary endocrine therapy (Wink et al., 2012). After a mean follow-up of 2.6 years, 58% achieved an initial response (with time to response of 7 months), 12% with stable disease, but 35% eventually had disease progression. Sixty-five percent of patients died during the follow-up period, of which only 14% died as a result of their BC. Compared to the historical 5-year overall survival rate of 62.3% of similarly aged patients treated with surgery ($\pm$ adjuvant pharmacologic therapy) during the same time period, the 5-year overall survival of these patients treated with primary endocrine therapy was only 27%. As compared to control patients, women treated with primary endocrine therapy were older and had more comorbid conditions compared to those treated with surgery.

Unfortunately, many of these randomized studies assessing the role of primary endocrine therapy in BC management have been largely restricted by small sample size and unselected for ER status (Johnston & Cheung, 2015). Additionally, there is some evidence showing a significant correlation between the degree of ER-positivity and response and outcome of patients receiving primary endocrine therapy (Johnston & Cheung, 2015). Although ongoing studies of primary endocrine therapy may reveal a particular subset of older adults who are fit for surgery who may be ultimately treated with primary endocrine therapy alone. A recent systemic review of clinical trials evaluating the effectiveness of endocrine therapy alone in women aged 70 years or older (with operable BC and fit for surgery) concluded that primary endocrine therapy should be reserved for those unfit for surgery or who refuse surgery (Morgan et al., 2014). Likewise, the authors concluded that long-term data suggests that many women eventually experience symptomatic progression, ultimately requiring more invasive interventions at a later age. Although the relative toxicity of primary endocrine therapy remains low compared to other treatment modalities, the definite role for adoption of these non-standard less aggressive therapies in older adults requires further study.

5.3.2 Chemotherapy

Historically, few women aged 70 years or older entered chemotherapy trials (EBCTCG et al., 2005). In fact, one study demonstrates that only 9% of patients enrolled in trials were $\geq$75 years of age, compared to 31% of the general patient population (Shachar et al., 2016). Prior randomized studies aiming to answer questions regarding elimination of chemotherapy in older women with BC have also terminated early due to insufficient accrual

(O'Connor et al., 2013). Together these factors make it difficult to fully ascertain the magnitude of benefit and risk of specific therapies, or lack thereof, in a large cohort of older women. However, older adults treated within the context of clinical trials experience similar treatment benefits as in younger patients (Older Adult Oncology, NCCN Guidelines, Version 2.2017). It is estimated that women with non-metastatic BC aged 50–65 years will achieve a 20% proportional reduction in the risk of recurrence and an 11% proportional reduction in the risk of death with adjuvant chemotherapy (Barginear et al., 2014). It is also well-established that undertreatment can be detrimental in all BC patients, regardless of age (Muss, 2010). Undertreatment may result from patient refusal or treatment discontinuation due to complications from another serious medical problem (Blair et al., 2016). Similarly, prognosis is significantly improved when clinicians comply with consensus treatment guidelines regarding treatment agents, schedule and dosing. Older women however are more likely to be treated with lower chemotherapy doses than younger women (Muss, 2010; Muss et al., 2009).

Although many older patients enrolled in clinical trials tolerate intensive regimens, in general, older adults remain vulnerable to increased chemotherapy toxicity due to associated comorbidities and underyling organ dysfunction (Barginear et al., 2014; Tesarova, 2013). Studies have evaluated the use of alternative chemotherapeutic regimens thought to be better tolerated in older adults. The phase 3 CALGB 49907 study, randomly assigning patients age ≥65 with stage I–III BC to standard adjuvant chemotherapy (cyclophosphamide, methotrexate, and fluorouracil or cyclophosphamide plus doxorubicin) vs capecitabine, found that standard chemotherapy regimens were associated with superior relapse-free survival to capecitabine (Muss, 2010). Likewise, in the randomized phase 3 ELDA study of women aged 65–79 with early stage BC at average to high risk of recurrence, adjuvant weekly docetaxel was associated with worse quality of life, severe non-hematological toxicity without being more effective than standard CMF (cyclophosphamide, methotrexate, fluorouracil) chemotherapy (Perrone et al., 2015). These studies highlight the importance of utilizing standard chemotherapy whenever possible, regardless of chronologic age. Tools, including the Cancer and Aging Research group chemotherapy toxicity calculator (http://www.mycarg.org/Chemo_Toxicity_Calculator, accessed September 2017) and the Moffitt Cancer Center CRASH (Chemotherapy Risk Assessment Scale for High-Age Patients) score (https://www.moffitt.org/eforms/crashscoreform, accessed September

2017) (Extermann et al., 2012; Hurria et al., 2016; Shachar et al., 2016) are available to objectively assess an older adult's risk of chemotherapy-related toxicity and can be helpful when planning systemic chemotherapy.

Genomic assays can also be helpful in determining the utility of chemotherapy in older adults, sparing those who are less likely to benefit. In the National Surgical Adjuvant Breast and Bowel Project (NSABP) B-14 clinical trial, the 21-gene RT-PCR assay used to determine the likelihood of distant recurrence by risk group in patients with node-negative BC treated with tamoxifen was found to be predictive independent of age ($P < 0.001$) (Paik et al., 2004). In one analysis of this assay, a greater proportion of younger patients had a high recurrence risk score than older patients, with increased likelihood of higher proportion of ER expression and lower recurrence score with advancing age (Swain et al., 2015). These data support the use of adjunct tools in the appropriate clinical context to aid in treatment decision-making of medically fit women with BC, some of whom may be safely spared the toxicity of unnecessary treatment. Nonetheless, a recently published large retrospective analysis of the National Cancer Database found that rates of non-testing increased with age in eligible patients (Kozick et al., 2018).

5.3.3 Anti-HER2 therapy

Although only a minority of older patients have been included in the pivotal studies of the HER2-directed therapies, subgroup outcomes analyses in this older cohort consistently demonstrate survival benefit when these standard of care therapies are utilized in management of HER2-positive BC. Such studies also suggest, however, that older patients are more vulnerable to the toxicities associated with these targeted therapies. Population-based data shows that increased cardiotoxicity following trastuzumab is seen with increasing age (Singh & Lichtman, 2015). Specifically, in a large series of women treated with trastuzumab, the incidence of cardiac dysfunction was 0.2% in women aged <50 compared to 10% in women aged >70 years, irrespective of cardiovascular status. The HR for cardiac toxicity was an astounding 11.3 for women above age 70 (Chavez-MacGregor et al., 2013). Older age (>80 years) and comorbidities, including obesity, hypertension, and coronary artery disease, have been identified as risk factors associated with congestive heart failure following trastuzumab in various analyses. Prior exposure to anthracyclines also augments risk for cardiac dysfunction following subsequent trastuzumab. In the large phase 3 CLEOPATRA (Clinical Evaluation of Pertuzumab and Trastuzumab) study, addition of pertuzumab to combined trastuzumab and docetaxel in the metastatic setting resulted in higher incidence of gastrointestinal (GI) toxicity in the older cohort (Miles et al., 2013; Swain et al., 2013). Interestingly, in this study of dual anti-HER2 blockade, age was not a statistically significant determinant of cardiac toxicity.

Collectively, these reports of use of anti-HER2 blockade in older women yet again support the concept that assessment of performance status, comorbidities and frailty through comprehensive geriatric assessments should dictate treatment inclusion rather than solely relying on chronologic age. Heightened awareness of the need to optimize cardiac function early on in the course of treatment to avert toxicity is imperative. This concept has heralded the emerging multidisciplinary field of cardio-oncology (Han et al., 2017). Strategies include an emphasis on diligent physical examination, routine cardiac monitoring as per consensus guidelines and incorporation of cardio-protective agents, such as beta blockers and/or angiotensin-converting-enzyme inhibitors (Singh & Lichtman, 2015).

5.4 Summary

Treatment decisions must not be based on chronologic age alone. Multiple studies have supported the notion that multi-modality medical and surgical treatments remain safe in older BC populations (Blair et al., 2016; Bouchardy et al., 2003; Vetter et al., 2013). Consideration of comorbid conditions, organ function and functional status is required to most objectively determine an individualized optimal treatment approach that will maintain a patient's quality of life. Likewise, expanding our experience with otherwise standard of care treatment options in older adults is critical to assuring safe and effective therapy in this patient cohort. For example, pooled outcomes data for patients with advanced hormone receptor positive BC aged ≥65 from three randomized studies of palbociclib and endocrine therapy found a higher incidence of myelosuppression in those aged ≥75 (without increase in grade 3 or higher myelosuppression among all age groups), but similar treatment benefit in this older population (Rugo et al., 2018). Currently, an ongoing phase 2 Alliance A1717601 study is assessing the tolerability and efficacy of palbociclib in combination with endocrine therapy (letrozole or fulvestrant) in patients with ER-positive, HER-2 positive metastatic BC over the age of 70 (https://www.ClinicalTrials.gov/, NCT03633331). Furthermore, investigating alternative treatment approaches in the context of a clinical trial is also critical in this patient population. For instance, a study is currently exploring the role of primary endocrine therapy in the management of early stage, potentially operable ER-positive, HER-2 negative BC with low 21-gene recurrence score and Ki-67/proliferative index $\leq$30% in older women aged $\geq$70 (https://www.ClinicalTrials.gov/, NCT02476786). An alternative study is exploring the incorporation of a decision aid on treatment options (benefits and risks of surgical and systemic procedures) in women aged $\geq$70 with ER-positive, HER2-negative, node-negative BC with consideration of a patient's competing comorbid conditions (https://www.ClinicalTrials.gov/, NCT02823262). The integration of standardized geriatric assessments and risk prediction tools into cancer care moves beyond an age-based approach to decision-making to help individualize selection of available locoregional and systemic treatment modalities that balance treatment efficacy, benefit and toxicity, and more accurately reflects a patient's biological age in the context of their predicted life expectancy.

6. Breast cancer survivorship in older women

Based on cancer incidence and estimated survival models from SEER program data, the largest increase in the total number of cancer survivors in 2020 is projected for those aged 65 years and older, with an expected increase of about 42% between 2010 and 2020 (Mariotto et al., 2011). Likewise, female BC is expected to have the largest number of all cancer survivors in 2020 (Mariotto et al., 2011).

In an effort to improve BC survivorship in older adults, formulation of comprehensive survivorship care plans that provide both education and highlight important aspects of ongoing cancer care have been strongly encouraged. These should highlight the complex needs of older adults, addressing issues pertaining to proper nutrition, physical mobility, resources for psychosocial support as well as management of comorbid conditions and potential polypharmacy (Mohile et al., 2016). Despite this particularly vulnerable population, only 35% of older women with early stage BC received a survivorship care plan in one study, with advancing age associated with lower likelihood of receiving a care plan (Mohile et al., 2016).

Many BC patients experience debility and fatigue following adjuvant BC therapy, particularly patients of advanced age following chemotherapy. Definitive data suggesting more prominent cancer- or cancer therapy-related fatigue in older women is lacking (VanderWalde & Hurria, 2012). Data, however, exists regarding the utility of yoga, sleep therapy and healthy diet in symptomatic management (VanderWalde & Hurria, 2012). Nevertheless, persistent decline in physical functioning beyond 1 year post BC diagnosis in older women has been shown to be associated with increased mortality, and is more commonly seen in patients with multiple comorbid conditions (Sehl et al., 2013). Additionally, cognitive impairment is not an uncommon complaint reported by BC patients following systemic anti-cancer therapies. Such impaired cognitive status post-treatment has been shown to be associated with worse survival, after adjusting for age, stage and cancer treatment. Comorbidity has been associated with pretreatment cognitive impairment in older BC patients (age >60) while not in controls, potentially suggesting increased susceptibility in patients with more comorbid conditions (Mandelblatt et al., 2014). Likewise, BC patients with advanced stages of disease tend to have lower executive function than those with early disease stage (Mandelblatt et al., 2014). Another study also found that age and baseline cognitive reserve seemed to be associated with rates of post-chemotherapy cognitive deficits, but these were short-term (VanderWalde & Hurria, 2012). This complication could impair a BC survivor's quality of life, mobility and adherence to treatment and survivorship care plan (Mandelbatt et al., 2003), but is difficult to accurately measure. Further studies of treatment-related cognitive impairment in older BC patients are needed to identify those at risk and determine preventative strategies.

7. Clinical trial considerations

Despite the increased prevalence of BC among women of advanced age, the exceedingly low accrual of older adults has been particularly notable in BC trials (Hutchins et al., 1999). Specifically, in a study of early stage BC patients, 34% of older patients were offered a clinical trial compared to 68% of younger patients (Kemeny et al., 2003). Additionally, a retrospective review of National Cancer Institute data involving 59,300 patients enrolled onto 495 cooperative group trials (1997–2000) found that only 32% of phase II–III trial participants were adults aged ≥65, and study exclusion criteria related to deficiencies in functional status and organ function were associated with lower enrollment of older adults (Lewis et al., 2003). Interestingly, studies suggest similar patient interest in trial participation from older adults as compared to younger counterparts (Kemeny et al., 2003). Stringent exclusion criteria often limit cancer clinical trial participation by older patients and those with other major chronic illnesses (Polite et al., 2017).

Even when older adults are included, they are limited to relatively healthy patients, thus limiting the applicability of studied interventions to most older patients. A large survey of BC providers' perceptions regarding barriers to clinical trial accrual of older BC patients found the most important factors were related to treatment-related toxicity and concerns about the ability to understand and comply with protocol procedures (Kornblith et al., 2002). Studies that address clinical questions of quality of life, treatment safety and efficacy relative to life expectancy are not only expected by our patients, but are also essential to streamlining treatment guidelines across healthcare systems to allow for evidence-based counseling by clinicians.

Several ongoing studies are addressing important considerations in older adults. One study is looking to establish the optimal strategy of fall-risk assessment to predict falls in older adults receiving systemic therapy for cancer (https://www.ClinicalTrials.gov/, NCT02912273, accessed September 2018). A phase 1 study is assessing the utility of a modified version of the Stepping On small group-based fall prevention educational program tailored to older adult patients in the oncology setting (https://www.ClinicalTrials.gov/, NCT02737839; http://ascopubs.org/doi/abs/10.1200/JCO.2018.36.15_suppl.e22020, accessed September 2018). Other studies are evaluating treatment-related side effects. One such study is assessing whether an educational program on self-management of chemotherapy-induced nausea and vomiting may help reduce cancer therapy-related complications in older patients (https://www.ClinicalTrials.gov/, NCT03143829, accessed September 2018). Another is investigating the efficacy of a telephone-administered cognitive-behavioral therapy intervention on anxiety, distress and quality of life outcomes in older adults (aged $\geq$65) with cancer and their caregiver (https://www.ClinicalTrials.gov/, NCT03168971, accessed September 2018).

This review highlights why we must invest both in the design of older adult-specific clinical trials (with thoughtful trial design and aims that focus on clinically meaningful issues to the geriatric population), and the recruitment of larger numbers of older adults into phase III BC trials on a global level. Studies should incorporate comprehensive geriatric assessments and treatment toxicity calculators to objectively risk stratify vulnerable cohorts to modifiable treatment algorithms, allowing for subsequent transition and applicability to routine clinical practice. Such collaborative strategies have been initiated by various groups, including the Cancer and Aging Research Group, National Cancer Institute and the National Institute on Aging (Barginear et al., 2014; Dale et al., 2012; Mohile et al., 2016).

8. Conclusion

Aging is a heterogeneous process that contributes to the challenging and complex nature of treatment decisions and the potential for therapy-related morbidity and mortality in older adults with BC. This can contribute to under- or overtreatment, thereby negatively impacting long-term outcomes. Based on the range of cognitive, physical and functional health status among older BC patients, age should not be the sole criterion for determining management of the older adult with BC. Multiple patient-specific factors must also play a role in treatment decision-making, including the patient and family preference as well as psychosocial issues. With the increasing complexity of cancer care, the integration of an interdisciplinary team, including geriatricians, palliative care specialists, pharmacists, therapists and dieticians, to name a few, may serve to further ensure that older patients with BC receive ideal cancer care and support. Likewise, various tools are currently available to assist oncologists in estimating life expectancy, screening patients for geriatric syndromes and predicting risk of chemotherapy toxicity in older women with BC. These tools allow for an individualized treatment approach and should be utilized in guiding both screening and treatment choices and counseling older women about their treatment options in the context of their physiologic and functional health status. Prioritizing focused and strategic research efforts will help narrow the disparities gap, improving the quality of care for those most likely to be affected by BC.

References

Arleo, E. K., Hendrick, R. E., Helvie, M. A., & Sickles, E. A. (2017). Comparison of recommendations for screening mammography using CISNET models. *Cancer, 123*(19), 3673–3680. https://doi.org/10.1002/cncr.30842.

Balducci, L., & Yates, J. (2000). General guidelines for the management of older patients with cancer. *Oncology, 14*(11A), 221–227.

Barginear, M. F., Muss, H., Kimmick, G., Owusu, C., Mrozek, E., Shahrokni, A., et al. (2014). Breast cancer and aging: Results of the U13 conference breast cancer panel. *Breast Cancer Research and Treatment, 146*(1), 1–6. https://doi.org/10.1007/s10549-014-2994-7.

Berry, D. A., Cronin, K. A., Plevritis, S. K., Fryback, D. G., Clarke, L., Zelen, M., et al. (2005). Effect of screening and adjuvant therapy on mortality from breast cancer. *The New England Journal of Medicine, 353*(17), 1784–1792. https://doi.org/10.1056/NEJMoa050518.

Blair, S., Robles, J., Weiss, A., Ward, E., & Unkart, J. (2016). Treatment of breast Cancer in women aged 80 and older: A systematic review. *Breast Cancer: Current Research, 1*, 115. https://doi.org/10.4172/2572-4118.1000115.

Bleyer, A., & Welch, H. G. (2012). Effect of three decades of screening mammography on breast-cancer incidence. *The New England Journal of Medicine, 367*(21), 1998–2005. https://doi.org/10.1056/NEJMoa1206809.

Botteri, E., Bagnardi, V., Goldhirsch, A., Viale, G., & Rotmensz, N. (2010). Axillary lymph node involvement in women with breast cancer: Does it depend on age? *Clinical Breast Cancer, 10*(4), 318–321. https://doi.org/10.3816/CBC.2010.n.042.

Bouchardy, C., Rapiti, E., Fioretta, G., Laissue, P., Neyroud-Caspar, I., Schäfer, P., et al. (2003). Undertreatment strongly decreases prognosis of breast cancer in elderly women. *Journal of Clinical Oncology, 21*(19), 3580–3587. https://doi.org/10.1200/JCO.2003.02.046.

Braithwaite, D., Demb, J., & Henderson, L. M. (2016). Optimal breast cancer screening strategies for older women: Current perspectives. *Clinical Interventions in Aging, 11*, 111–125. https://doi.org/10.2147/CIA.S65304.

Butz, D. R., Lapin, B., Yao, K., Wang, E., Song, D. H., Johnson, D., et al. (2015). Advanced age is a predictor of 30-day complications after autologous but not implant-based postmastectomy breast reconstruction. *Plastic and Reconstructive Surgery, 135*(2), 253e–261e. https://doi.org/10.1097/PRS.0000000000000988.

Chavez-MacGregor, M., Zhang, N., Buchholz, T. A., Zhang, Y., Niu, J., Elting, L., et al. (2013). Trastuzumab-related cardiotoxicity among older patients with breast cancer. *Journal of Clinical Oncology, 31*(33), 4222–4228. https://doi.org/10.1200/JCO.2013.48.7884.

Chen, H. L., Zhou, M. Q., Tian, W., Meng, K. X., & He, H. F. (2016). Effect of age on breast Cancer patient prognoses: A population-based study using the SEER 18 database. *PLoS One, 11*(10), e0165409. https://doi.org/10.1371/journal.pone.0165409.

Chirgwin, J. H., Giobbie-Hurder, A., Coates, A. S., Price, K. N., Ejlertsen, B., Debled, M., et al. (2016). Treatment adherence and its impact on disease-free survival in the breast international group 1-98 trial of tamoxifen and Letrozole, alone and in sequence. *Journal of Clinical Oncology, 34*(21), 2452–2459. https://doi.org/10.1200/JCO.2015.63.8619. See comment in PubMed Commons below.

Chlebowski, R. T., Haque, R., Hedlin, H., Col, N., Paskett, E., Manson, J. E., et al. (2015). Benefit/risk for adjuvant breast cancer therapy with tamoxifen or aromatase inhibitor use by age, and race/ethnicity. *Breast Cancer Research and Treatment, 154*(3), 609–616. https://doi.org/10.1007/s10549-015-3647-1.

Chow, W. B., Rosenthal, R. A., Merkow, R. P., Ko, C. Y., Esnaola, N. F., & American College of Surgeons National Surgical Quality Improvement Program; American Geriatrics Society. (2012). Optimal preoperative assessment of the geriatric surgical patient: A best practices guideline from the American College of Surgeons National Surgical Quality Improvement Program and the American Geriatrics Society. *Journal of the American College of Surgeons, 215*(4), 453–466. https://doi.org/10.1016/j.jamcollsurg.2012.06.017.

Dale, W., Mohile, S. G., Eldadah, B. A., Trimble, E. L., Schilsky, R. L., Cohen, H. J., et al. (2012). Biological, clinical, and psychosocial correlates at the interface of cancer and aging research. *Journal of the National Cancer Institute, 104*(8), 581–589. https://doi.org/10.1093/jnci/djs145.

Daugherty, E. C., Daugherty, M. R., Bogart, J. A., & Shapiro, A. (2016). Adjuvant radiation improves survival in older women following breast-conserving surgery for estrogen receptor-negative breast Cancer. *Clinical Breast Cancer, 16*(6), 500–506.e2. https://doi.org/10.1016/j.clbc.2016.06.017.

Early Breast Cancer Trialists' Collaborative Group (EBCTCG). (2005). Effects of chemotherapy and hormonal therapy for early breast cancer on recurrence and 15-year survival: An overview of the randomised trials. *Lancet, 365*(9472), 1687–1717. https://doi.org/10.1016/S0140-6736(05)66544-0.

Early Breast Cancer Trialists' Collaborative Group (EBCTCG). Darby, S., McGale, P., Correa, C., Taylor, C., Arriagada, R., et al. (2011). Effect of radiotherapy after breast-conserving surgery on 10-year recurrence and 15-year breast cancer death: Meta-analysis of individual patient data for 10,801 women in 17 randomised trials. *Lancet, 378*(9804), 1707–1716. https://doi.org/10.1016/S0140-6736(11)61629-2.

Early Breast Cancer Trialists' Collaborative Group (EBCTCG). Peto, R., Davies, C., Godwin, J., Gray, R., Pan, H. C., et al. (2012). Comparisons between different polychemotherapy regimens for early breast cancer: Meta-analyses of long-term outcome among 100,000 women in 123 randomized trials. *Lancet, 379*(9814), 432–444. https://doi.org/10.1016/S0140-6736(11)61625-5.

Eaton, B. R., Jiang, R., Torres, M. A., Kahn, S. T., Godette, K., Lash, T. L., et al. (2016). Benefit of adjuvant radiotherapy after breast-conserving therapy among elderly women with T1-T2N0 estrogen receptor-negative breast cancer. *Cancer, 122*(19), 3059–3068. https://doi.org/10.1002/cncr.30142.

Extermann, M., Boler, I., Reich, R. R., Lyman, G. H., Brown, R. H., DeFelice, J., et al. (2012). Predicting the risk of chemotherapy toxicity in older patients: The chemotherapy risk assessment scale for high-age patients (CRASH) score. *Cancer, 118*(13), 3377–3386. https://doi.org/10.1002/cncr.26646.

Fischer, J. P., Wes, A. M., Tuggle, C. T., & Wu, L. C. (2014). Venous thromboembolism risk in mastectomy and immediate breast reconstruction: Analysis of the 2005 to 2011 American College of Surgeons National Surgical Quality Improvement Program data sets. *Plastic and Reconstructive Surgery, 133*(3), 263e–273e. https://doi.org/10.1097/01.prs.0000438062.53914.22.

Freedman, R. A., Keating, N. L., Pace, L. E., Lii, J., McCarthy, E. P., & Schonberg, M. A. (2017). Use of surveillance mammography among older breast cancer survivors by life expectancy. *Journal of Clinical Oncology, 35*(27), 3123–3130. https://doi.org/10.1200/JCO.2016.72.1209.

Freedman, R. A., Keating, N. L., Partridge, A. H., Muss, H. B., Hurria, A., & Winer, E. P. (2017). Surveillance mammography in older patients with breast cancer—Can we ever stop? A review. *JAMA Oncology, 3*(3), 402–409. https://doi.org/10.1001/jamaoncol.2016.3931.

Freedman, R. A., & Partridge, A. H. (2017). Emerging data and current challenges for young, old, obese, or male patients with breast cancer. *Clinical Cancer Research, 23*(11), 2647–2654. https://doi.org/10.1158/1078-0432.

Gennari, R., Curigliano, G., Rotmensz, N., Robertson, C., Colleoni, M., Zurrida, S., et al. (2004). Breast carcinoma in elderly women: Features of disease presentation, choice of local and systemic treatments compared with younger postmenopasual patients. *Cancer, 101*(6), 1302–1310. https://doi.org/10.1002/cncr.20535.

Graetz, I., McKillop, C. N., Stepanski, E., Vidal, G. A., Anderson, J. N., & Schwartzberg. (2018). Use of a web-based app to improve breast cancer symptom management and adherence for aromatase inhibitors: A randomized controlled feasibility trial. *Journal of Cancer Survivorship, 12*(4), 431–440. https://doi.org/10.1007/s11764-018-0682-z.

Güth, U., Myrick, M. E., Kandler, C., & Vetter, M. (2013). The use of adjuvant endocrine breast cancer therapy in the oldest old. *Breast, 22*(5), 863–868. https://doi.org/10.1016/j.breast.2013.03.001.

Hamaker, M. E., Seynaeve, C., Wymenga, A. N., van Tinteren, H., Nortier, J. W., Maartense, E., et al. (2014). Baseline comprehensive geriatric assessment is associated with toxicity and survival in elderly metastatic breast cancer patients receiving single-agent chemotherapy: Results from the OMEGA study of the Dutch breast cancer trialists' group. *Breast, 23*(1), 81–87. https://doi.org/10.1016/j.breast.2013.11.004.

Han, X., Zhou, Y., & Liu, W. (2017). Precision cardio-oncology: Understanding the cardiotoxicity of cancer therapy. *Nature Partner Journal (npj) Precision Oncology*, *1*(1), 31. https://doi.org/10.1038/s41698-017-0034-x. eCollection 2017.

Henderson, L. M., O'Meara, E. S., Braithwaite, D., & Onega, T. (2015). Performance of digital screening mammography among older women in the United States. *Cancer*, *121*(9), 1379–1386. https://doi.org/10.1002/cncr.29214.

Hershman, D. L., Kushi, L. H., Shao, T., Buono, D., Kershenbaum, A., Tsai, W. Y., et al. (2010). Early discontinuation and nonadherence to adjuvant hormonal therapy in a cohort of 8,769 early-stage breast cancer patients. *Journal of Clinical Oncology*, *28*(27), 4120–4128. https://doi.org/10.1200/JCO.2009.25.9655.

Hind, D., Wyld, L., Beverley, C. B., & Reed, M. W. (2006). Surgery versus primary endocrine therapy for operable primary breast cancer in elderly women (70 years plus). *Cochrane Database of Systematic Reviews*, (1). CD004272. https://doi.org/10.1002/14651858.CD004272.pub2.

Hughes, K. S., Schnaper, L. A., Bellon, J. R., Cirrincione, C. T., Berry, D. A., McCormick, B., et al. (2013). Lumpectomy plus tamoxifen with or without irradiation in women age 70 years or older with early breast cancer: Long-term follow-up of CALGB 9343. *Journal of Clinical Oncology*, *31*(19), 2382–2387. https://doi.org/10.1200/JCO.2012.45.2615. See comment in PubMed Commons below.

Hughes, K. S., Schnaper, L. A., Berry, D., Cirrincione, C., McCormick, B., Shank, B., et al. (2004). Lumpectomy plus tamoxifen with or without irradiation in women 70 years of age or older with early breast cancer. *The New England Journal of Medicine*, *351*(10), 971–977. https://doi.org/10.1056/NEJMoa040587.

Huisingh-Scheetz, M., & Walston, J. (2017). How should older adults with cancer be evaluated for frailty? *Journal of Geriatric Oncology*, *8*(1), 8–15. https://doi.org/10.1016/j.jgo.2016.06.003.

Hurria, A., Leung, D., Trainor, K., Borgen, P., Norton, L., & Hudis, C. (2003). Factors influencing treatment patterns of breast cancer patients age 75 and older. *Critical Reviews in Oncology/Hematology*, *46*(2), 121–126.

Hurria, A., Mohile, S., Gajra, A., Klepin, H., Muss, H., Chapman, A., et al. (2016). Validation of a prediction tool for chemotherapy toxicity in older adults with Cancer. *Journal of Clinical Oncology*, *34*(20), 2366–2371. https://doi.org/10.1200/JCO.2015.65.4327.

Hutchins, L. F., Unger, J. M., Crowley, J. J., Coltman, C. A., Jr., & Albain, K. S. (1999). Underrepresentation of patients 65 years of age or older in cancer-treatment trials. *The New England Journal of Medicine*, *341*(27), 2061–2067. https://doi.org/10.1056/NEJM199912303412706.

International Breast Cancer Study Group, Rudenstam, C. M., Zahrieh, D., Forbes, J. F., Crivellari, D., Holmberg, S. B., et al. (2006). Randomized trial comparing axillary clearance versus no axillary clearance in older patients with breast cancer: First results of International Breast Cancer Study Group Trial 10-93. *Journal of Clinical Oncology*, *24*(3), 337–344. https://doi.org/10.1200/JCO.2005.01.5784.

Janssen-Heijnen, M. L., Houterman, S., Lemmens, V. E., Louwman, M. W., Maas, H. A., & Coebergh, J. W. (2005). Prognostic impact of increasing age and co-morbidity in cancer patients: A population-based approach. *Critical Reviews in Oncology/Hematology*, *55*(3), 231–240. https://doi.org/10.1016/j.critrevonc.2005.04.008.

Jiang, M., Hughes, D. R., Appleton, C. M., McGinty, F., & Duszak, R. (2015). Recent trends in adherence to continuous screening for breast cancer among Medicare beneficiaries. *Preventive Medicine*, *73*, 47–52. https://doi.org/10.1016/j.ypmed.2014.12.031.

Johnston, S. J., & Cheung, K. L. (2015). The role of primary endocrine therapy in older women with operable breast cancer. *Future Oncology*, *11*(10), 1555–1565. https://doi.org/10.2217/fon.15.13.

Kantor, O., Pesce, C., Liederbach, E., Wang, C. H., Winchester, D. J., & Yao, K. (2016). Surgery and hormone therapy trends in octogenarians with invasive breast cancer. *American Journal of Surgery, 211*(3), 541–545. https://doi.org/10.1016/j.amjsurg.2015.11.005.

Keating, N. L., Landrum, M. B., Guadagnoli, E., Winer, E. P., & Ayanian, J. Z. (2006). Factors related to underuse of surveillance mammography among breast cancer survivors. *Journal of Clinical Oncology, 24*(1), 85–94. https://doi.org/10.1200/JCO.2005.02.4174.

Kemeny, M. M., Peterson, B. L., Kornblith, A. B., Muss, H. B., Wheeler, J., Levine, E., et al. (2003). Barriers to clinical trial participation by older women with breast cancer. *Journal of Clinical Oncology, 21*(12), 2268–2275. https://doi.org/10.1200/JCO.2003.09.124.

Kirkhus, L., Šaltytė Benth, J., Rostoft, S., Grønberg, B. H., Hjermstad, M. J., Selbæk, G., et al. (2017). Geriatric assessment is superior to oncologists' clinical judgement in identifying frailty. *British Journal of Cancer, 117*(4), 470–477. https://doi.org/10.1038/bjc.2017.202.

Königsberg, R., Pfeiler, G., Hammerschmid, N., Holub, O., Glossmann, K., Larcher-Senn, J., et al. (2016). Breast cancer subtypes in patients aged 70 years and older. *Cancer Investigation, 34*(5), 197–204. https://doi.org/10.1080/07357907.2016.1182184.

Kornblith, A. B., Kemeny, M., Peterson, B. L., Wheeler, J., Crawford, J., Bartlett, N., et al. (2002). Survey of oncologists' perceptions of barriers to accrual of older patients with breast carcinoma to clinical trials. *Cancer, 95*(5), 989–996. https://doi.org/10.1002/cncr.10792.

Kozick, Z., Hashmi, A., Dove, J., Hunsinger, M., Arora, T., Wild, J., et al. (2018). Disparities in compliance with the Oncotype DX breast cancer test in the United States: A National Cancer Data Base assessment. *American Journal of Surgery, 215*(4), 686–692. https://doi.org/10.1016/j.amjsurg.2017.05.008. Epub 2017 Jun 14.

Kuijer, A., & King, T. A. (2017). Age, molecular subtypes and local therapy decision-making. *Breast, 34*(Suppl. 1), S70–S77. https://doi.org/10.1016/j.breast.2017.06.032.

Kunkler, I. H., Williams, L. J., Jack, W. J., Cameron, D. A., Dixon, J. M., & Prime II investigators. (2015). Breast-conserving surgery with or without irradiation in women aged 65 years or older with early breast cancer (PRIME II): A randomized controlled trial. *The Lancet Oncology, 16*(3), 266–273. https://doi.org/10.1016/S1470-2045(14)71221-5.

Le Saux, O., Ripamonti, B., Bruyas, A., Bonin, O., Freyer, G., Bonnefoy, M., et al. (2015). Optimal management of breast cancer in the elderly patient: Current perspectives. *Clinical Interventions in Aging, 10*, 157–174. https://doi.org/10.2147/CIA.S50670.

Lewis, J. H., Kilgore, M. L., Goldman, D. P., Trimble, E. L., Kaplan, R., Montello, M. J., et al. (2003). Participation of patients 65 years of age or older in cancer clinical trials. *Journal of Clinical Oncology, 21*(7), 1383–1389. https://doi.org/10.1200/JCO.2003.08.010.

Løberg, M., Lousdal, M. L., Bretthauer, M., & Kalager, M. (2015). Benefits and harms of mammography screening. *Breast Cancer Research, 17*, 63. https://doi.org/10.1186/s13058-015-0525-z.

Maly, R. C., Liu, Y., Kwong, E., Thind, A., & Diamant, A. L. (2009). Patient-physician communication in breast reconstructive surgery. *Cancer, 115*(20), 4819–4827. https://doi.org/10.1002/cncr.24510.

Mandelblatt, J., Figueiredo, M., & Cullen, J. (2003). Outcomes and quality of life following breast cancer treatment in older women: When, why, how much, and what do women want? *Health and Quality of Life Outcomes, 1*, 45. https://doi.org/10.1186/1477-7525-1-45.

Mandelblatt, J. S., Hadley, J., Kerner, J. F., Schulman, K. A., Gold, K., Dunmore-Griffith, J., et al. (2000). Patterns of breast carcinoma treatment in older women: Patient preference and clinical and physical influences. *Cancer, 89*(3), 561–573.

Mandelblatt, J. S., Stern, R. A., Luta, G., & McGuckin, M. (2014). Cognitive impairment in older patients with breast cancer before systemic therapy: Is there an interaction between cancer and comorbidity? *Journal of Clinical Oncology*, *32*(18), 1909–1918. https://doi.org/10.1200/JCO.2013.54.2050.

Mariotto, A. B., Etzioni, R., Hurlbert, M., Penberthy, L., & Mayer, M. (2017). Estimation of the number of women living with metastatic breast cancer in the United States. *Cancer Epidemiology, Biomarkers & Prevention*, *26*(6), 809–815. https://doi.org/10.1158/1055-9965.

Mariotto, A. B., Yabroff, K. R., Shao, Y., Feuer, E. J., & Brown, M. L. (2011). Projections of the cost of cancer care in the United States: 2010-2020. *Journal of the National Cancer Institute*, *103*(2), 117–128. https://doi.org/10.1093/jnci/djq495.

Markopoulos, C., & van de Water, W. (2012). Older patients with breast cancer: Is there bias in the treatment they receive? *Therapeutic Advances in Medical Oncology*, *4*(6), 321–327. https://doi.org/10.1177/1758834012455684.

Miles, D., Baselga, J., Amadori, D., Sunpaweravong, P., Semiglazov, V., Knott, A., et al. (2013). Treatment of older patients with HER2-positive metastatic breast cancer with pertuzumab, trastuzumab, and docetaxel: Subgroup analyses from a randomized, double-blind, placebo-controlled phase III trial (CLEOPATRA). *Breast Cancer Research and Treatment*, *142*(1), 89–99. https://doi.org/10.1007/s10549-013-2710-z.

Mohile, S. G., Hurria, A., Cohen, H. J., Rowland, J. H., Leach, C. R., Arora, N. K., et al. (2016). Improving the quality of survivorship for older adults with cancer. *Cancer*, *122*(16), 2459–2568. https://doi.org/10.1002/cncr.30053.

Morgan, J., Wyld, L., Collins, K. A., & Reed, M. W. (2014). Surgery versus primary endocrine therapy for operable primary breast cancer in elderly women (70 years plus). *Cochrane Breast Cancer Group*. https://doi.org/10.1002/14651858.CD004272.pub3.

Muss, H. B. (2010). Coming of age: Breast cancer in seniors. *The Oncologist*, *15*(Suppl. 5), 57–65. https://doi.org/10.1634/theoncologist.2010-S5-57.

Muss, H. B., Berry, D. A., Cirrincione, C. T., Theodoulou, M., Mauer, A. M., Kornblith, A. B., et al. (2009). Adjuvant chemotherapy in older women with early-stage breast cancer. *The New England Journal of Medicine*, *360*(20), 2055–2065. https://doi.org/10.1056/NEJMoa0810266.

Muss, H. B., Tu, D., Ingle, J. N., Martino, S., Robert, N. J., Pater, J. L., et al. (2008). Efficacy, toxicity, and quality of life in older women with early-stage breast cancer treated with letrozole or placebo after 5 years of tamoxifen: NCIC CTG intergroup trial MA.17. *Journal of Clinical Oncology*, *26*(12), 1956–1964. https://doi.org/10.1200/JCO.2007.12.6334.

Nabholtz, J. A. (2008). Long-term safety of aromatase inhibitors in the treatment of breast cancer. *Therapeutics and Clinical Risk Management*, *4*(1), 189–204.

Nagar, H., Yan, W., Christos, P., Chao, K. S. C., Nori, D., & Ravi, A. (2017). Older patients wih early-stage breast cancer: Adjuvant radiation therapy and predictive factors for cancer-related death. *American Journal of Clinical Oncology*, *40*(3), 300–305. https://doi.org/10.1097/COC.0000000000000144.

NCCN. (2017). *Clinical Practice Guidelines in Oncology: Older Adult Oncology. Version 2. May 1, 2017. https://www.nccn.org/professionals/physician_gls/pdf/senior.pdf.*

Nelson, H. D., O'Meara, E. S., Kerlikowske, K., Balch, S., & Miglioretti, D. (2016). Factors associated with rates of false-positive and false-negative results from digital mammography screening: An analysis of registry data. *Annals of Internal Medicine*, *164*(4), 226–235. https://doi.org/10.7326/M15-0971.

Nelson HD, Tyne K, Naik A, Bougatsos C, Chan BK, Humphrey L (2009). U.S. Preventive Services Task Force. Screening for breast cancer: An update for the U.S. Preventive Services Task Force. Annals of Internal Medicine 151 (10), 727–37, W237-42. doi: https://doi.org/10.7326/0003-4819-151-10-200911170-00009.

O'Connor, T., Shinde, A., Doan, C., Katheria, V., & Hurria, A. (2013). Managing breast cancer in the older patient. *Clinical Advances in Hematology & Oncology*, *11*(6), 341–347.

O'Donnell, S., Goldstein, B., Dimatteo, M. R., Fox, S. A., John, C. R., & Obrzut, J. E. (2010). Adherence to mammography and colorectal cancer screening in women 50-80 years of age the role of psychological distress. *Womens Health Issues*, *20*(5), 343–349. https://doi.org/10.1016/j.whi.2010.04.002.

Oeffinger, K. C., Fontham, E. T., Etzioni, R., Herzig, A., Michaelson, J. S., Shih, Y. C., et al. (2015). Breast cancer screening for women at average risk: 2015 guideline update from the American Cancer Society. *Journal of the American Medical Association*, *314*(15), 1599–1614. https://doi.org/10.1001/jama.2015.12783.

Ogle, K. S., Swanson, G. M., Woods, N., & Azzouz, F. (2000). Cancer and comorbidity: Redefining chronic diseases. *Cancer*, *88*(3), 653–663.

Oh, D. D., Flitcroft, K., Brennan, M. E., Snook, K. L., & Spillane, A. (2018). Patient-reported outcomes of breast reconstruction in older women: Audit of a large metropolitan public/private practice in Sydney, Australia. *Psychooncology*, *27*(12), 2815–2822. https://doi.org/10.1002/pon.4895. Epub 2018 Oct 10.

Oh, D. D., Flitcroft, K., Brennan, M. E., & Spillane, A. J. (2016). Patterns and outcomes of breast reconstruction in older women—A systematic review of the literature. *European Journal of Surgical Oncology*, *242*(5), 604–615. https://doi.org/10.1016/j.ejso.2016.02.010.

Ong, C. T., Thomas, S. M., Blitzblau, R. C., Fayanju, O. M., Park, T. S., Plichta, J. K., et al. (2017). Patient age and tumor subtype predict the extent of axillary surgery among breast cancer patients eligible for the American College of Surgeons Oncology Group Trial Z0011. *Annals of Surgical Oncology*, *24*(12), 3559–3566. https://doi.org/10.1245/s10434-017-6075-0. Epub 2017 Sep 6.

Orucevic, A., Curzon, M., Curzon, C., Heidel, R. E., McLoughlin, J. M., Panella, T., et al. (2015). Breast cancer in elderly Caucasian women-an institution-based study of correlation between breast cancer prognostic markers, TNM stage, and overall survival. *Cancers (Basel)*, 7(3), 1472–1483. https://doi.org/10.3390/cancers7030846.

Osborn, G., Jones, M., Champ, C., Gower-Thomas, K., & Vaughan-Williams, E. (2011). Is primary endocrine therapy effective in treating the elderly, unfit patient with breast cancer? *Annals of the Royal College of Surgeons of England*, *93*(4), 286–289. https://doi.org/10.1308/003588411X571917.

Paik, S., Shak, S., Tang, G., Kim, C., Baker, J., Cronin, M., et al. (2004). A multigene assay to predict recurrence of tamoxifen-treated, node-negative breast cancer. *The New England Journal of Medicine*, *351*(27), 2817–2826. https://doi.org/10.1056/NEJMoa041588.

Patnaik, J. L., Byers, T., Diguiseppi, C., Denberg, T. D., & Dabelea, D. (2011). The influence of comorbidities on overall survival among older women diagnosed with breast cancer. *Journal of the National Cancer Institute*, *103*, 1101–1111. https://doi.org/10.1093/jnci/djr188.

Perrone, F., Nuzzo, F., Di Rella, F., Gravina, A., Iodice, G., Labonia, V., et al. (2015). Weekly docetaxel versus CMF as adjuvant chemotherapy for older women with early breast cancer: Final results of the randomized phase III ELDA trial. *Annals of Oncology*, *26*(4), 675–682. https://doi.org/10.1093/annonc/mdu564.

Polite, B. N., Adams-Campbell, L. L., Brawley, O. W., Bickell, N., Carethers, J. M., Flowers, C. R., et al. (2017). Charting the future of cancer health disparities research: A position statement from the American Association for Cancer Research, the American Cancer Society, the American Society of Clinical Oncology, and the National Cancer Institute. *Journal of Clinical Oncology*, *35*(26), 3075–3082. https://doi.org/10.1200/JCO.2017.73.6546.

Ring, A., Reed, M., Leonard, R., Kunkler, I., Muss, H., Wildiers, H., et al. (2011). The treatment of early breast cancer in women over the age of 70. *British Journal of Cancer, 105*, 189–193. https://doi.org/10.1038/bjc.2011.234.
Rugo, H. S., Turner, N. C., Finn, R. S., Joy, A. A., Verma, S., Harbeck, N., et al. (2018). Palbociclib plus endocrine therapy in older women with HR+/HER2- advanced breast cancer: A pooled analysis of randomised PALOMA clinical studies. *European Journal of Cancer, 101*, 123–133. https://doi.org/10.1016/j.ejca.2018.05.017.
Santosa, K. B., Qi, J., Kim, H. M., Hamill, J. B., Pusic, A. L., & Wilkins, E. G. (2016). Effect of patient age on outcomes in breast reconstruction: Results from a multicenter prospective study. *Journal of the American College of Surgeons, 2223*(6), 745–754. https://doi.org/10.1016/j.jamcollsurg.2016.09.003.
Schneider, R., Barakat, A., Pippen, J., & Osborne, C. (2011). Aromatase inhibitors in the treatment of breast cancer in post-menopausal female patients: An update. *Breast Cancer (Dove Med Press), 3*, 113–125. https://doi.org/10.2147/BCTT.S22905.
Schonberg, M. A. (2016). Decision-making regarding mammography screening for older women. *Journal of the American Geriatrics Society, 64*(12), 2413–2418. https://doi.org/10.1111/jgs.14503.
Schonberg, M. A., Marcantonio, E. R., Li, D., Silliman, R. A., Ngo, L., & McCarthy, E. P. (2010). Breast cancer among the oldest old: Tumor characteristics, treatment choices, and survival. *Journal of Clinical Oncology, 28*(12), 2038–2045. https://doi.org/10.1200/JCO.2009.25.9796. See comment in PubMed Commons below.
Schonberg, M. A., Silliman, R. A., McCarthy, E. P., & Marcantonio, E. R. (2014). Factors noted to affect women aged 80 and older's breast cancer treatment decisions. *Breast Cancer Research and Treatment, 145*(1), 211–223. https://doi.org/10.1007/s10549-014-2921-y.
Searle, S. D., Mitnitksi, A., Gahbauer, E. A., Gill, T. M., & Rockwood, K. (2008). A standard procedure for creating a frailty index. *BMC Geriatrics, 8*, 24. https://doi.org/10.1186/1471-2318-8-24.
Sehl, M., Lu, X., Silliman, R., & Ganz, P. A. (2013). Decline in physical functioning in first 2 years after breast cancer diagnosis predicts 10-year survival in older women. *Journal of Cancer Survivorship*, 7(1), 20–31. https://doi.org/10.1007/s11764-012-0239-5.
Shachar, S. S., Hurria, A., & Muss, H. B. (2016). Breast Cancer in women older than 80 years. *Journal of Oncology Practice, 12*(2), 123–132. https://doi.org/10.1200/JOP.2015.010207.
Sheppard, V. B., Faul, L. A., Luta, G., Clapp, J. D., Yung, R. L., Wang, J. H., et al. (2014). Frailty and adherence to adjuvant hormonal therapy in older women with breast cancer: CALGB protocol 369901. *Journal of Clinical Oncology, 32*(22), 2318–2327. https://doi.org/10.1200/JCO.2013.51.7367See comment in PubMed Commons below.
Singh, R., Hellman, S., & Heimann, R. (2004). The natural history of breast carcinoma in the elderly—Implications for screening and treatment. *Cancer, 100*, 1807–1813. https://doi.org/10.1002/cncr.20206.
Singh, J. C., & Lichtman, S. M. (2015). Targeted agents for HER2-positive breast cancer: Optimal use in older patients. *Drugs & Aging, 32*(12), 975–982. https://doi.org/10.1007/s40266-015-0326-1.
Smith, B. D., Jiang, J., McLaughlin, S. S., Hurria, A., Smith, G. L., Giordano, S. H., et al. (2011). Improvement in breast cancer outcomes over time: Are older women missing out? *Journal of Clinical Oncology, 29*(35), 4647–4653. https://doi.org/10.1200/JCO.2011.35.8408.
Swain, S. M., Kim, S. B., Cortés, J., Ro, J., Semiglazov, V., Campone, M., et al. (2013). Pertuzumab, trastuzumab, and docetaxel for HER2-positive metastatic breast cancer (CLEOPATRA study): Overall survival results from a randomised, double-blind, placebo-controlled, phase 3 study. *The Lancet Oncology, 14*(6), 461–471. https://doi.org/10.1016/S1470-2045(13)70130-X.

Swain, S. M., Nunes, R., Yoshizawa, C., Rothney, M., & Sing, A. P. (2015). Quantitative gene expression by recurrence score in ER-positive breast cancer, by age. *Advances in Therapy*, *32*(12), 1222–1236. https://doi.org/10.1007/s12325-015-0268-3.

Tesarova, P. (2013). Breast cancer in the elderly—Should it be treated differently? *Reports of Practical Oncology and Radiotherapy*, *18*(1), 26–33. https://doi.org/10.1016/j.rpor.2012.05.005.

Tesarova, P. (2016). Specific aspects of breast cancer therapy of elderly women. *BioMed Research International*, *2016*, 1381695. https://doi.org/10.1155/2016/1381695.

Turner, N., Zafarana, E., Becheri, D., Mottino, G., & Biganzoli, L. (2013). Breast cancer in the elderly: Which lessons have we learned? *Future Oncology*, *9*(12), 1871–1881. https://doi.org/10.2217/fon.13.140.

US Preventative Services Task Force (USPSTF). (2009). Screening for breast cancer: U.S. preventative services task force recommendation statement. *Annals of Internal Medicine*, *151*(10), 716–726. https://doi.org/10.7326/0003-4819-151-10-200911170-00008.

VanderWalde, A., & Hurria, A. (2012). Early breast cancer in the older woman. *Clinics in Geriatric Medicine*, *28*(1), 73–91. https://doi.org/10.1016/j.cger.2011.10.002.

Vetter, M., Huang, D. J., Bosshard, G., & Guth, U. (2013). Breast cancer in women 80 years of age and older: A comprehensive analysis of an underreported entity. *Acta Oncologica*, *52*(1), 57–65. https://doi.org/10.3109/0284186X.2012.731523.

Walter, L. C., & Schonberg, M. A. (2014). Screening mammography in older women: A review. *JAMA*, *311*(13), 1336–1347. https://doi.org/10.1001/jama.2014.2834.

Wildiers, H., Kunkler, I., Biganzoli, L., Fracheboud, J., Vlastos, G., Bernard-Marty, C., et al. (2007). Management of breast cancer in elderly individuals: Recommendations of the International Society of Geriatric Oncology. *The Lancet Oncology*, *8*(12), 1101–1115. https://doi.org/10.1016/S1470-2045(07)70378-9.

Wildiers, H., Van Calster, B., van de Poll-Franse, L. V., Hendrickx, W., Røislien, J., Smeets, A., et al. (2009). Relationship between age and axillary lymph node involvement in women with breast cancer. *Journal of Clinical Oncology*, *27*(18), 2931–2937. https://doi.org/10.1200/JCO.2008.16.7619.

Williams, G. R., Mackenzie, A., Magnuson, A., Olin, R., Chapman, A., Mohile, S., et al. (2016). Comorbidity in older adults with cancer. *Journal of Geriatric Oncology*, 7(4), 249–257. https://doi.org/10.1016/j.jgo.2015.12.002.

Wink, C. J., Woensdregt, K., Nieuwenhuijzen, G. A., van der Sangen, M. J., Hutschemaekers, S., Roukema, J. A., et al. (2012). Hormone treatment without surgery for patients aged 75 years or older with operable breast cancer. *Annals of Surgical Oncology*, *19*(4), 1185–1191. https://doi.org/10.1245/s10434-011-2070-z.

Yancik, R. (1997). Cancer burden in the aged: An epidemiologic and demographic overview. *Cancer*, *80*(7), 1273–1283.

Yancik, R., Wesley, M. N., Ries, L. A., Havlik, R. J., Edwards, B. K., & Yates, J. W. (2001). Effect of age and comorbidity in postmenopausal breast cancer patients aged 55 years and older. *JAMA*, *285*(7), 885–892. https://doi.org/10.1001/jama.285.7.885.

Yang, P., Du, C. W., Kwan, M., Liang, S. X., & Zhang, G. J. (2013). The impact of p53 in predicting clinical outcome of breast cancer patients with visceral metastasis. *Scientific Reports*, *3*, 2246. https://doi.org/10.1038/srep02246.

Zhu, W., Perez, E. A., Hong, R., Li, Q., & Xu, B. (2015). Age-related disparity in immediate prognosis of patients with triple-negative breast cancer: A population-based study from SEER cancer registries. *PLoS One*, *10*(5), e0128345. https://doi.org/10.1371/journal.pone.0128345. eCollection 2015.

Further reading

Bergman, L., Kluck, H. M., van Leeuwen, F. E., Crommelin, M. A., Dekker, G., Hart, A. A., et al. (1992). The influence of age on treatment choice and survival of elderly breast cancer patients in South-Eastern Netherlands: A population-based study. *European Journal of Cancer, 28A*(8–9), 1475–1480.

Biganzoli, L., Wildiers, H., Oakman, C., Marotti, L., Loibl, S., Kunkler, I., et al. (2012). Management of elderly patients with breast cancer: Updated recommendations of the International Society of Geriatric Oncology (SIOG) and European Society of Breast Cancer Specialists (EUSOMA). *The Lancet Oncology, 13*(4), e148–e160. https://doi.org/10.1016/S1470-2045(11)70383-7.

Clinical Treatment Score post 5 years. CTS5 calculator. https://www.cts5-calculator.com/.

Demb, J., Allen, I., & Braithwaite, D. (2016). Utilization of screening mammography in older women according to comorbidity and age: Protocol for a systematic review. *Systematic Reviews, 5*(1), 168. https://doi.org/10.1186/s13643-016-0345-y.

Giuliano, A. E., Hunt, K. K., Ballman, K. V., Beitsch, P. D., Whitworth, P. W., Blumencranz, P. W., et al. (2011). Axillary dissection vs no axillary dissection in women with invasive breast cancer and sentinel node metastasis: A randomized clinical trial. *JAMA, 305*, 569e75. https://doi.org/10.1001/jama.2011.90.

Greer, L. T., Rosman, M., Charles Mylander, W., Liang, W., Buras, R. R., Chagpar, A. B., et al. (2014). A prediction model for the presence of axillary lymph node involvement in women with invasive breast cancer: A focus on older women. *The Breast Journal, 20*(2), 147–153. https://doi.org/10.1111/tbj.12233.

Hind, D., Wyld, L., & Reed, M. W. (2007). Surgery, with or without tamoxifen, vs tamoxifen alone for older women with operable breast cancer: Cochrane review. *British Journal of Cancer, 96*(7), 1025–1029. https://doi.org/10.1038/sj.bjc.6603600.

Louwman, W. J., Vulto, J. C., Verhoeven, R. H., Nieuwenhuijzen, G. A., Coebergh, J. W., & Voogd, A. C. (2007). Clinical epidemiology of breast cancer in the elderly. *European Journal of Cancer, 43*(15), 2242–2252. https://doi.org/10.1016/j.ejca.2007.08.005.

Mandelblatt, J. S., Cai, L., Luta, G., Kimmick, G., Clapp, J., Isaacs, C., et al. (2017). Frailty and long-term mortality of older breast cancer patients: CALGB 369901 (Alliance). *Breast Cancer Research and Treatment, 164*(1), 107–117. https://doi.org/10.1007/s10549-017-4222-8.

Murphy, C. C., Bartholomew, L. K., Carpentier, M. Y., Bluethmann, S. M., & Vernon, S. W. (2012). Adherence to adjuvant hormonal therapy among breast cancer survivors in clinical practice: A systematic review. *Breast Cancer Research and Treatment, 134*(2), 459–478. https://doi.org/10.1007/s10549-012-2114-5.

National Cancer Institute Surveillance, n.d. "Epidemiology and End Results: Breast cancer incidence and mortality," http://seer.cancer.gov/statfacts/html/breast.html.

NCT02476786. Endocrine Treatment Alone for Elderly Patients With Estrogen Receptor Positive Operable Breast Cancer and Low Recurrence Score. https://clinicaltrials.gov/ct2/show/NCT02476786. Accessed September 2018 n.d.

NCT02737839. Adaptation of the STEPPING ON Fall Prevention Program for Older Adults Receiving Cancer Therapy. https://clinicaltrials.gov/ct2/show/NCT02737839. Accessed September 2018 n.d.

NCT02823262. A Breast Cancer Treatment Decision Aid for Women Aged 70 and Older. https://clinicaltrials.gov/ct2/show/NCT02823262. Accessed September 2018 n.d.

NCT02912273. Optimizing Fall-risk Prediction in Older Adults With Cancer. https://clinicaltrials.gov/ct2/show/NCT02912273. Accessed September 2018 n.d.

NCT03143829 Promoting Cancer Symptom Management in Older Adults. https://clinicaltrials.gov/ct2/show/NCT03143829. Accessed September 2018 n.d.

NCT03168971. Managing Anxiety From Cancer (MAC): Testing an Intervention for Anxiety in Older Adults With Cancer and Their Caregivers. https://clinicaltrials.gov/ct2/show/NCT03168971. Accessed September 2018 n.d.

NCT03633331. Palbociclib and Letrozole or Fulvestrant in Treating Patients With Estrogen Receptor Positive, HER2 Negative Metastatic Breast Cancer https://clinicaltrials.gov/ct2/show/NCT03633331. Accessed September 2018 n.d.

Social Security Administration. (). *Retirement and Survivors Benefits Life Expectancy Calculator. https://www.ssa.gov/OACT/population/longevity.html.*

Walter, L. C., Eng, C., & Covinsky, K. E. (2001). Screening mammography for frail older women: What are the burdens? *Journal of General Internal Medicine, 16*, 779–784. https://doi.org/10.1111/j.1525-1497.2001.10113.x.

CHAPTER THREE

Pubertal mammary development as a "susceptibility window" for breast cancer disparity

Bradley Krisanits, Jaime F. Randise, Clare E. Burton, Victoria J. Findlay, David P. Turner*

Department of Pathology & Laboratory Medicine, Hollings Cancer Center, Medical University of South Carolina, Charleston, SC, United States

*Corresponding author: e-mail address: turnerda@musc.edu

Contents

Abstract

Factors such as socioeconomic status, age at menarche and childbearing patterns are components that have been shown to influence mammary gland development and establish breast cancer disparity. Pubertal mammary gland development is selected as the focus of this review, as it is identified as a "window of susceptibility" for breast cancer risk and disparity. Here we recognize non-Hispanic White, African American, and Asian American women as the focus of breast cancer disparity, in conjunction with diets associated with changes in breast cancer risk. Diets consisting of high fat, N-3

Advances in Cancer Research, Volume 146
ISSN 0065-230X
https://doi.org/10.1016/bs.acr.2020.01.004

polyunsaturated fatty acids, N-6 polyunsaturated fatty acids, as well as obesity and the Western diet have shown to lead to changes in pubertal mammary gland development in mammalian models, therefore increasing the risk of breast cancer and breast cancer disparity. While limited intervention strategies are offered to adolescents to mitigate development changes and breast cancer risk, the prominent solution to closing the disparity among the selected population is to foster lifestyle changes that avoid the deleterious effects of unhealthy diets.

1. Introduction

The mammary gland is one of the few organs that continues to develop postnatally through stages of massive tissue remodeling that is directly influenced and controlled by hormonal signaling. This includes stages of development involving proliferation, differentiation and apoptosis. These stages of development also include multiple cell types including stromal and epithelial crosstalk (Inman et al., 2015; Macias & Hinck, 2012; Sternlicht, 2006; Sternlicht et al., 2006). These timepoints of tightly regulated development are now being described as "windows of susceptibility" where modifiable risk factors, including diet, obesity, hormone therapy, alcohol consumption, and smoking can negatively influence normal mammary gland regulation and thereby increase the risk of developing breast cancer (Isaac, Sundar, Romanos, & Rahman, 2016; Martinson et al., 2013; Okasha et al., 2003; Sundaram, Johnson, & Makowski, 2013).

Studies have shown that specific populations around the world are at a greater risk for developing breast cancer. Breast cancer is the most prevalent female cancer accounting for nearly one in three cancers diagnosed in this population. Unfortunately, although recent statistics show that incidence rates are similar, mortality rates are over 40% higher in African American (AA) women compared to Non-Hispanic White (NHW) Women (Churpek et al., 2015; Ramnitz & Lodish, 2013). In addition, AA women are more likely to be diagnosed at later stages and to have the lowest survival rate at each stage of diagnosis. Several studies have shown that AA adolescents often begin puberty at younger ages than other races (Churpek et al., 2015). Menarche is defined as the first menstrual cycle in female humans. From both social and medical perspectives, it is often considered the central event of female puberty, as it signals the possibility of fertility. Females experience menarche at different ages.

The timing of menarche is influenced by female biology, as well as genetic and environmental factors, specifically nutritional factors. The mean age of menarche has declined over the last century, but the magnitude of the decline and the factors responsible remain subjects of contention. The worldwide average age of menarche is very difficult to estimate accurately, and it varies significantly by geographical region, race, ethnicity and other characteristics, but various estimates have placed it at 13 years of age. This is particularly relevant here, as the risk for developing breast cancer increases for each year a female starts menarche sooner. It is estimated that for each 2-year increase in actual age at menarche, there is a 10% reduction in breast cancer risk (Ambrosone et al., 2014; Carwile et al., 2015; Cui et al., 2014; Deardorff et al., 2014; Ramnitz & Lodish, 2013; Reagan et al., 2012). Despite this, the link between cancer disparity and the "windows of susceptibility" during mammary development have not been adequately assessed.

While intervention strategies during puberty have been cited as having the greatest impact on reducing breast cancer risk, the idea of prescribing young females with medication to prevent the future development of breast cancer is, for obvious reasons, controversial. However, current research is beginning to uncover alternative interventions, including diet, exercise, and other modifiable lifestyle factors as more appropriate and feasible strategies to decrease breast cancer risk in all women but in particular AA women who suffer from many of the risk factors that are independently associated with earlier menarche (Biro & Deardorff, 2013; Colditz & Frazier, 1995; Dieli-Conwright, Lee, & Kiwata, 2016; Ramnitz & Lodish, 2013; Reagan et al., 2012). This review will summarize the biological, socioeconomic and environmental associations between mammary development and breast cancer disparity rates.

2. Mammary gland development

The mammary gland, as mentioned previously, is a unique organ as it develops postnatally. Following pre-natal or embryonic development where a rudimentary ductal tree is formed, pubertal development is the next major stage of development to occur whereby the ductal tree is laid down in order for the mature functioning gland. This allows for the future development changes that occur during the cyclic stages of pregnancy, lactation and involution (Fig. 1). The mammary gland is a complex secretory

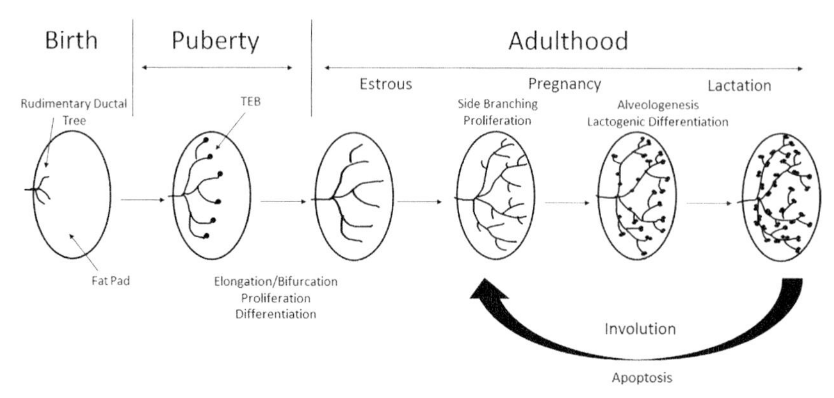

Fig. 1 Overview of different stages of mammary gland development. Schematic representation of mammary gland development from a rudimentary structure at birth, through tightly regulated phases of growth such as puberty, associated with the formation of TEBs and increased proliferation, differentiation and elongation/branching of the ductal tree. After puberty the gland is fully formed and becomes quiescent, until pregnancy. Hormonal factors during pregnancy leads to increased proliferation, differentiation, side branching and alveologenesis to form a functioning gland for lactation. Following lactation is a stage of tightly controlled apoptosis during involution to return the gland to a quiescent state.

organ composed of multiple cell types, including epithelial cells, adipocytes, endothelial cells, fibroblasts, and various immune cells (Brady, Chuntova, & Schwertfeger, 2016; Inman et al., 2015; Macias & Hinck, 2012; Sternlicht, 2006; Sternlicht et al., 2006). The main function of the mammary gland is to produce and secrete milk for the nourishment of newborns during breastfeeding and lactation. The gland fulfills its function of supplying adequate milk by forming an extensive and functional tree-like structure of branched ducts that fill the fat pad during pubertal development (Inman et al., 2015; Macias & Hinck, 2012; Sternlicht, 2006; Sternlicht et al., 2006). Reproductive development is primarily under hormonal influence leading to an increase in secondary and tertiary branching, vascularization, as well as alveologenesis during pregnancy, which later differentiate into milk-secreting lobules during lactation. After weaning of offspring, stagnate milk in the lobules sends local cues leading to cellular signaling and an immune response, which helps to guide the ductal tree back to a simple ductal architecture where it no longer produces or secretes milk (Inman et al., 2015; Macias & Hinck, 2012; Sternlicht, 2006; Sternlicht et al., 2006). This stage of development is called involution. The consequences of these tightly regulated phases of mammary development leads to periods that are vulnerable to cancer development, known as windows of susceptibility (Biro & Deardorff, 2013; Colditz & Frazier, 1995; Dieli-Conwright et al., 2016; Martinson et al., 2013; Sundaram et al., 2013).

2.1 Window of susceptibility

The "lifecycle" of mammary gland development can be divided into five windows of cancer susceptibility: in utero, pubertal, pregnancy, postpartum involution, and age-related involution. Fortunately, each of these windows shows a limited duration making the at-risk populations identifiable (Biro & Deardorff, 2013; Colditz & Frazier, 1995). A key characteristic of these risk "windows" is tightly regulated tissue remodeling driven by hormonal signaling, in addition to mammary epithelium and stromal crosstalk (Martinson et al., 2013; Sundaram et al., 2013). Epidemiological studies have established a connection between specific life events and breast cancer risk. Having a child later in life, as well as a child with high birth weight, increases breast cancer risk for woman in adulthood (Biro & Deardorff, 2013; Okasha et al., 2003; Reagan et al., 2012). Nutrition during adolescence has been shown to affect breast cancer risk, as increased consumption of milk, soy, fruits and vegetables, as well as caloric restriction lead to a decreased risk of breast cancer (Biro & Deardorff, 2013; Okasha et al., 2003; Shrivastava, Shrivastava, & Ramasamy, 2016). On the contra, increased consumption of visible fats of meat led to increased risk of premenopausal breast cancer (Biro & Deardorff, 2013; Okasha et al., 2003; Shrivastava et al., 2016). Non-dietary factors, such as physical activity during adolescence have also been shown to lead to a decrease in breast cancer risk (Biro & Deardorff, 2013; Dieli-Conwright et al., 2016; Okasha et al., 2003). Other specific life events including age at puberty, length of time between menarche and first birth, exposure to smoking and age at menopause have also been shown to affect breast cancer risk (Biro & Deardorff, 2013; Deardorff et al., 2014; Okasha et al., 2003). The recognition of these "windows of susceptibility" led to the proposal of targeting intervention strategies for cancer prevention at crucial stages of mammary gland development (Biro & Deardorff, 2013; Colditz & Frazier, 1995; Dieli-Conwright et al., 2016).

2.2 The pubertal window

The pubertal window is attractive to study as preventative measures at this stage have the potential to have the biggest impact on reducing future risk of breast disease. Unfortunately, it remains one of the least attractive for clinical translation due to the controversy surrounding the use of chemoprevention drug use in vulnerable populations. However, a greater understanding of the development of the mammary gland during puberty will lead to more attractive methods of prevention in this population. One of the best models that researchers use to study mammary development is rodents (mouse and rat). Indeed, research in rodent models has led to a better understanding of mammary development and has been applied to humans.

Pubertal development of the rodent mammary gland begins during an extensive hormone-dependent phase of growth associated with increased epithelial cell proliferation (Inman et al., 2015; Macias & Hinck, 2012; Sternlicht, 2006). This stage is associated with the formation of terminal end buds (TEBs), highly proliferative club-like structures at the ends of ducts (Ball, 1998; Butner et al., 2016; Paine & Lewis, 2017). These bulbous structures form around 3 weeks of age and begin to penetrate the fat pad via the proliferation of a single layer of cap cells at the ends of the TEBs and the underlying epithelium (Ball, 1998; Butner et al., 2016; Paine & Lewis, 2017). Cap cells then differentiate into the myoepithelial layer surrounding the tubular ductal bilayer as the tree pushes into the mammary fat pad (Ball, 1998; Butner et al., 2016; Paine & Lewis, 2017; Sternlicht, 2006; Sternlicht et al., 2006). Proliferation is regulated by growth hormones (GH), thereby inducing the expression of insulin-like growth factor-1 (IGF1) in the mammary stromal cells (Fig. 2). Acting together with estrogen (E2) from the ovaries, IGF1 helps to induce epithelial proliferation. Acting via a paracrine fashion, E2 signals through its receptor, estrogen receptor 1 (ER), to stimulate the release of amphiregulin (AREG), an epidermal growth factor (EGF) family member (Sternlicht, 2006; Sternlicht et al., 2006). AREG then binds to its receptor on stromal cells inducing the release of fibroblast growth factors (FGFs), which stimulate ER negative luminal cell proliferation (Macias & Hinck, 2012; Sternlicht, 2006).

Extensive primary ductal networks develop via the bifurcation of the TEBs followed by the divergence of secondary side branches. Secondary branching from primary ducts continues until the ductal tree occupies approximately 60% of the fat pad, leaving space for the influence of pregnancy hormones. The mature ductal tree undergoes further branching under the influence of cycling ovarian hormones, leading to the formation of short tertiary side branches (Inman et al., 2015; Macias & Hinck, 2012; Sternlicht, 2006; Sternlicht et al., 2006). Factors including TGF-β1, Reelin, Slit2 and Netrin1 lead to either positive or negative regulation of mammary ductal

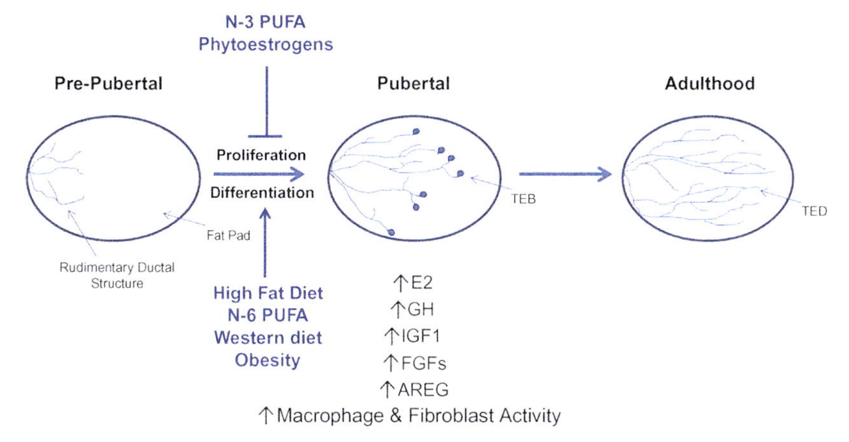

Fig. 2 Factors influencing pubertal mammary gland development. The tightly regulated window of pubertal mammary gland development has shown to be a key window of "cancer susceptibility" specific when looking at lifestyle factors, such as diet and obesity. Diets such as; a diet high in fat, N-6 PUFA, the Western diet and lifestyle factors such as obesity have been shown in various in vivo models to lead to dysregulation of development and increased hormonal and growth factor status as well as immune/stromal cell activity. Meanwhile other diets high in N-3 PUFA and phytoestrogens have been shown to have protective effects in cancer susceptibility when consumed during puberty.

formation (Macias & Hinck, 2012; Sternlicht, 2006; Sternlicht et al., 2006). Once the gland is fully formed at the end of pubertal development, highly proliferative TEBs begin to taper off and become quiescent forming terminal end ducts (TEDs) in adulthood (Ball, 1998; Butner et al., 2016; Paine & Lewis, 2017).

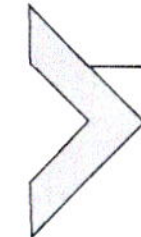

3. Mammary gland development and cancer health disparity

Breast cancer is the most commonly diagnosed cancer and is the second leading cause of cancer related mortality in women. Although breast cancer incidence is equal among AA and NHW women, AA women are more likely to die of their disease. In fact, more aggressive tumor characteristics are more common in breast cancers diagnosed in AA women than any other racial/ethnic group. Cancer stage at diagnosis directly affects overall survival, and only 52% of breast cancer cases in AA women are diagnosed at local stage, compared to 63% in NHW women. In addition, 22% of breast cancers in AA women are classified as triple negative compared to only

10–12% in women of other racial groups. Since there are currently no targeted therapies for this aggressive subtype of breast cancer it remains the deadliest with the poorest prognosis (Ambrosone et al., 2014). AA women are also more likely to have ER −, PR − and double negative (ER −/PR −) breast cancer compared to NHW women, which is associated with higher grade, later stage and reduced survival rate (Cui et al., 2014). Several biological, socioeconomic and environmental factors exist between mammary development and breast cancer disparity rates, these will be discussed below.

3.1 Socioeconomic and environmental factors

The disparity observed in breast cancer survival outcomes may be attributed in part to unhealthy lifestyle and socioeconomic factors, as well as factors associated with access to sufficient medical care. Studies show that as many as 70% of new breast cancer cases may be attributed to diet and poor lifestyle (Willett, 1999). Dietary factors, in particular, are known to affect mammary gland development during puberty and contribute to increased tumor occurrence and progression, discussed in more detail in Section 4 (Shrivastava et al., 2016; Song et al., 2014). Low-income populations are associated with a poor diet consisting of low cost, unhealthy, and highly processed foods. This contributes to the disparities observed as AA communities have the highest prevalence of low-income with over 27% living at or below the poverty level. These populations are also 1.5 times more likely to be obese compared to NHW populations (Economic Policy Institute, n.d.; U.S. Department of Health and Human Service, n.d.). Low-income, poor diet, lack of exercise, and certain life choices such as smoking and drinking reveal a socioeconomic and environmental connection between mammary development, breast cancer risk, and cancer disparity. Historically, a person of a lower socioeconomic status is more likely to integrate themselves with factors that increase breast cancer risk. Marketing strategies, environmental and community factors, such as reduced access to fruits and vegetables and fewer opportunities for physical activity, contribute to the integration of low socioeconomic status and breast cancer risk.

3.2 Menarche

Age at menarche is defined as the age of first menstrual cycle in females and has been linked with increased risk of breast cancer development for each year earlier than average (12.77 years in the United States) at which it occurs. It is estimated that for each 2-year increase in age at menarche, there is a 10% reduction in breast cancer risk. Studies have shown that both AA women and Hispanic women are more likely to have earlier onset of menarche compared to NHW women (Ambrosone et al., 2014; Carwile et al., 2015; Cui et al., 2014; Deardorff et al., 2014; Ramnitz & Lodish, 2013; Reagan et al., 2012). Age at menarche was shown to occur earlier in populations associated with increased poverty levels, and as discussed earlier poverty levels are higher in AA populations (Reagan et al., 2012). NHW women who experienced a later menarche were shown to have a decreased risk for the development of solely ER positive (ER +) breast cancer when compared to AA women, which showed a decreased risk of both ER +/− breast cancer when menarche was later (Ambrosone et al., 2014; Cui et al., 2014). A plethora of factors are thought to cause changes in menarche onset, such as: social economic status (lower family education level and lower family income), obesity during childhood, physical inactivity, and diet (Carwile et al., 2015; Cui et al., 2014; Deardorff et al., 2014; Mervish et al., 2017; Mueller et al., 2015; Ramnitz & Lodish, 2013; Reagan et al., 2012; Vani et al., 2013; Villamor & Jansen, 2016).

3.3 Childbearing patterns

Childbearing patterns have also been shown to differ by race/ethnicity, with higher parity (defined as the number of times a woman has carried a pregnancy to a viable gestational age) and a lower prevalence of lactation in AA women, both of which have been associated with an increased risk of breast cancer (Ambrosone et al., 2014; Cui et al., 2014; Deardorff et al., 2014). In NHW women, it was observed that multiparity (≥ 3) led to a decreased risk of ER + breast cancer risk, while late age of first birth (> 30 years) and nulliparity led to an increased risk of the development of ER + breast cancer (Cui et al., 2014). A woman's risk of breast cancer decreases by 7% for each birth, further signifying the effect parity has on breast cancer prevention. Shorter interval between menarche and first live birth has also been shown to lead to an increased risk of ER negative (ER −) breast cancer (Ambrosone et al., 2014). This is noteworthy, as the current trends are both earlier age at menarche and later age of first birth, both of which serve to increase this interval. Breastfeeding is the optimal method to deliver nutrients to newborns, but also has plentiful benefits on both the infant and the mother. The World Health Organization and American Academy of Pediatrics recommends exclusive breastfeeding for the first 6 months of an infant's life and continuing breastfeeding up to 2 years (World Health Organization, 2017). Breastfeeding has also been shown to affect a woman's breast cancer risk. It has been suggested that the benefits of breastfeeding surpass previously mentioned childbearing patterns, including parity (Phipps & Li, 2014). Furthermore, breastfeeding is a modifiable risk factor which is important for focusing on cancer prevention (Anstey, Chen, Elam-Evans, & Perrine, 2017; Phipps & Li, 2014). Indeed several studies have now shown that women who breastfeed for 4–6 months show a 25–50% reduction in breast cancer risk compared to parous women that have never breastfed (Phipps & Li, 2014). This was further emphasized in a 2002 study that demonstrated a 4.3% decrease in breast cancer risk for every 12 months a mother breastfed (Collaborative Group on Hormonal Factors in Breast Cancer, 2002). Interestingly, there is a significant difference in breastfeeding practices among AA women; data obtained from the Black Women's Health Study (BWHS; $n = 35{,}338$ black) and the Nurses' Health Study II (NHSII; $n = 103{,}508$ white and 2068 black) showed that the age-standardized prevalence of ever breastfeeding was only 44% of multiparous AA women, compared to NHW women who reported 81% (Phipps & Li, 2014; Warner et al., 2013). Data collected in these prospective cohorts were by self-reported questionnaires. This is a continuing trend, as recent national statistics reported only 58.9% of new mothers of AA descent reported breastfeeding, compared to 75.2% of new mothers of NHW descent (Anstey et al., 2017; Phipps & Li, 2014; Smith et al., 2005). These data were collected as part of the National Immunization Survey (NIS) by person to person interview in households with children aged 19–35 months (Smith et al., 2005). One study showed that if AA women breastfed at the same rate as NHW women, the incidence of triple-negative breast cancer in the United States would be reduced by two-thirds among parous AA women (Phipps & Li, 2014). Many factors contribute to this disparity within the AA community and include a lack of information provided about the benefits of breastfeeding, a lack of resources to assist, and interestingly, a lack of social and cultural acceptance within the AA community toward breastfeeding (Anstey et al., 2017). Differences in breastfeeding practices can be contributed to differences in social economic status, as well as more biological factors like obesity, development, diet, physical activity or other chronic diseases (Anstey et al., 2017; Dieli-Conwright et al., 2016; Phipps & Li, 2014).

4. Diet, development and breast cancer risk

Many studies have examined the impact of diet on development and breast cancer risk and these will be discussed in detail below. This information is also summarized in Table 1 and Fig. 2.

4.1 High-fat diet

A high-fat diet (HFD) is a diet consisting of at least 35% of total calories is consumed from fats, both unsaturated and saturated. In addition to the popular processed foods, many other foods have a high fat content including but not limited to animal fat, chocolate, butter, and oily fish. Commonly higher in fat content, most processed foods are easier to obtain as they are normally cheaper considering socioeconomical factors, such as lower family income. Many dishes among different cultures and ethnicities such as fried foods or "soul food" contain ingredients with high fat such as oils, butters, and fats to increase flavor and appeal.

Animal studies have explored the impact of high-fat diet on pubertal growth in various strains of mice. Genetic background is important as many studies have shown that different outcomes can be observed depending on the strain used. C57BL/6 mice fed a HFD during puberty led to reduced ductal length, as well as number of TEBs, a sparse ductal tree, increased mammary adiposity, and reduced mammary epithelial cell proliferation when compared to mice fed a control diet (Olson et al., 2010). A HFD in C57BL/6 mice was also shown to reduce E2 responsiveness, a key hormonal factor in the developing mammary gland. When BALB/c mice were fed a HFD during puberty, a similar morphological change was observed in the glands; however, BALB/c mice showed increased mammary epithelial cell proliferation and reduced E2 responsiveness with no change in adiposity of the gland. BALB/c mice, genetically related to A/J mice, are less susceptible to HFD-induced obesity compared to C57BL/6 as seen in the increased mammary adiposity and associated differences among the strains. Interestingly, weight loss initiated in C57BL/6 mice from switching of diets, HFD to control, restored TEB formation and ductal elongation, showing weight gain and mammary gland adiposity to be key players in pubertal mammary development (MacLennan & Ma, 2010; Olson et al., 2010). BALB/c mice fed a HFD also showed increased recruitment of immune cells, such as eosinophils and mast cells, to periepithelial mammary stroma,

Table 1 Overview of studies investigating the impact of diet on development and breast cancer risk.

Environmental factor	Developmental effect	Breast cancer risk
High-fat diet	↑Mammary adiposity (C57BL/6) (Warner et al., 2013) ↓Epithelial proliferation (C57BL/6) (Warner et al., 2013) ↑Epithelial proliferation (BALB/c) (Smith et al., 2005) ↓E2 responsiveness (BALB/c) (Smith et al., 2005) ↓Duct length and number of TEBs (Warner et al., 2013) ↑Vascularization and immune cell recruitment (Olson et al., 2010) Sparse ductal tree (Warner et al., 2013)	↓Latency of DMBA/PhIP-induced mammary tumors (Olson et al., 2010; Zhao et al., 2013; Zhu et al., 2016) ↑Growth factors, stromal invasion, inflammatory and angiogenic processes (Aupperlee et al., 2015; MacLennan & Ma, 2010) ↑Proliferation, tumor growth, M2 macrophage recruitment and vascularization (Aupperlee et al., 2015; MacLennan & Ma, 2010) Intervention after puberty, reduced tumor latency remained (Aupperlee et al., 2015) Loss of luminal gene expression (Snyderwine et al., 1998) ↑Mesenchymal and breast cancer invasive gene expression (Snyderwine et al., 1998)
Phytoestrogens	↑Acceleration of mammary cell differentiation (Shao, 1998; Yu, Zhang, & Wu, 2003) ↑Proliferation and number of TEBs in early development (Shao, 1998; Yu et al., 2003)	Abundant in blood and urine of low breast cancer populations (Brzezinski & Debi, 1999; Jawaid, Crane, Nowers, Lacey, & Whitehead, 2010) Associated with later age of menarche (Brzezinski & Debi, 1999; Mervish et al., 2017) ↑Progesterone, estrogen receptor expression and proliferation of lobular epithelium (Iwasaki et al., 2008) ↓HER2 and neu expression (Iwasaki et al., 2008) Inhibit growth of estrogen dependent and independent breast cancer cell lines (Anderson et al., 2014; Maskarinec, Verheus, & Tice, 2010; Wu et al., 2002) ↓Chemically-induced mammary tumors in mice fed pubertal diet Antioxidant effect (Shao, 1998; Wu et al., 2002; Yu et al., 2003) ↓Angiogenesis, EGF signaling pathway and local synthesis of estrogen (Maskarinec et al., 2010; Shao, 1998; Wiseman et al., 2000; Yu et al., 2003)

Table 1 Overview of studies investigating the impact of diet on development and breast cancer risk.—cont'd

Environmental factor	Developmental effect	Breast cancer risk
N-3 Polyunsaturated fatty acids	Delay in pubertal onset (Abdelmagid et al., 2016; Lamartiniere, Murril, et al., 1998; Lamartiniere, Zhang, & Cotroneo, 1998; Wu et al., 2002) ↓TEBs in early development (Abdelmagid et al., 2016; Lamartiniere, Murril, et al., 1998; Lamartiniere, Zhang, et al., 1998; Wu et al., 2002) ↓TEB proliferation (Abdelmagid et al., 2016) ↑TEB apoptosis, ductal coverage and elongation (Abdelmagid et al., 2016; Lamartiniere, Zhang, et al., 1998; Wu et al., 2002)	Abundant in populations with low breast cancer incidence (Manni et al., 2011; Willett, 1999; Zhu et al., 2011) Delay in pubertal onset (Abdelmagid et al., 2016; Lamartiniere, Murril, et al., 1998; Lamartiniere, Zhang, et al., 1998; Wu et al., 2002) Anti-inflammatory and modulation of oncogenic signaling via lipid raft disruption (Bagga et al., 2002; Goodstine et al., 2003; Lamartiniere, Zhang, et al., 1998; Leslie et al., 2014; Saadatian-Elahi et al., 2004; Wu et al., 2002; Zhu et al., 2011) ↑Apoptosis, expression of proteins in cell cycle control and DNA repair (Bagga et al., 2002; Goodstine et al., 2003; Lamartiniere, Zhang, et al., 1998; Leslie et al., 2014; Saadatian-Elahi et al., 2004; Wu et al., 2002; Zhu et al., 2011) ↓Cellular proliferation and mammographic density (Lamartiniere, Murril, et al., 1998; Lamartiniere, Zhang, et al., 1998; Liu & Ma, 2014; Luijten et al., 2013; MacLennan et al., 2013; Medvedovic et al., 2009; Olivo & Hilakivi-Clarke, 2005; Saadatian-Elahi et al., 2004; Weisburger, 1997; Wu et al., 2002) Inhibit the effects of oncogenic N-6 PUFAs (Goodstine et al., 2003; Luijten et al., 2013)
N-6 Polyunsaturated fatty acids	Early pubertal onset (Lamartiniere, Zhang, et al., 1998; Wu et al., 2002) ↑TEBs in early development (Lamartiniere, Zhang, et al., 1998; Wu et al., 2002) ↑Proliferation and AA in glands (Lamartiniere, Zhang, et al., 1998; Wu et al., 2002)	↑Expression of human breast cancer genes (Zhu et al., 2011) ↑AA leading to precursors active in carcinogenesis (Goodstine et al., 2003; Lamartiniere, Zhang, et al., 1998; Zhu et al., 2011) Unbalanced ratio of N-6/N-3 leads to inflammation, modulation of oncogenic protein signaling and increased proliferation (Diorio & Dumas, 2014; Goodstine et al., 2003; Lamartiniere, Zhang, et al., 1998; Liu & Ma, 2014; Luijten et al., 2013)

Continued

Table 1 Overview of studies investigating the impact of diet on development and breast cancer risk.—cont'd

Environmental factor	Developmental effect	Breast cancer risk
Western diet	↑Proliferation of epithelial cells of TEBs (Escrich et al., 2004; Khan et al., 1994; Uribarri et al., 2010) ↑TEB and duct presence (Escrich et al., 2004; Khan et al., 1994; Uribarri et al., 2010)	Exposure to diet after immigration leads to loss of protective nature (Brzezinski & Debi, 1999; Jawaid et al., 2010; Messina, Caan, Abrams, Hardy, & Maskarinec, 2013; Wiseman et al., 2000) Early menarche onset (Carwile et al., 2015; Mueller et al., 2015; Villamor & Jansen, 2016) ↑Consumption of reactive compounds leading to manipulation of cell cycle and survival (Fritz et al., 2013; Snyderwine, 1998; Xue et al., 1996)
Obesity	Enlarged mammary gland (Gooderham et al., 2002) Sparse ductal tree, surrounded by thick collagen layers and incomplete myoepithelium layer (Gooderham et al., 2002) ↓Ductal branching (Gooderham et al., 2002)	↑Inflammation and formation of CLS (Kamikawa et al., 2009; Lauber & Gooderham, 2011) ↑Pro-inflammatory factors and aromatase (Kamikawa et al., 2009; Lauber & Gooderham, 2011) ↑NF-κβ, hypertrophy of adipocytes and hormone receptors on tumors (Bhardwaj et al., 2015; Kamikawa et al., 2009; Lauber & Gooderham, 2011; Subbaramaiah et al., 2011)

as well as hyperplastic lesions during pubertal development (3 weeks). Increased vascularization was observed later in pubertal development to sustain the increased proliferation (Aupperlee et al., 2015).

Rodent models have also explored the impact of HFD on breast cancer risk. BALB/c mice fed a HFD throughout puberty, showed a reduced latency of a median time of 115 days versus 204 in LFD in 7,12-dimethylbenz(a)anthracene (DMBA)-induced mammary tumors, which were similar to human basal-like breast cancer (Aupperlee et al., 2015). The group showed that the reduced latency is most likely a result of increased growth factor expression, as well as increased inflammatory and angiogenic processes. Prior to tumor formation, mice fed the HFD showed increased proliferation, hyperplasia, and macrophage recruitment. Resultant tumors also showed increased proliferation, M2 macrophage recruitment, as well as increased vascularization. Interestingly, mice fed a HFD diet during early puberty (3 weeks), then switched to a low-fat diet (LFD) in late puberty (9 weeks), still showed similar reduced latency in tumors with human basal-like characteristics compared to mice fed only HFD for 45 weeks, emphasizing the importance of the pubertal window of insult (Zhao et al., 2013; Zhu et al., 2016). Similar studies in rats showed that treatment with the carcinogen 2-amino-1-methyl-6-phenylimidazo [4,5-*b*]pyridine (PhIP) during early puberty (~6 weeks of age) and fed a HFD for 25 weeks, results in an increase in tumor incidence with associated increases in tumor proliferation and growth, as well as stromal invasion compared to rats fed a LFD (Snyderwine, Davis, Schut, & Roberts-Thomson, 1998; Snyderwine et al., 1998). Furthermore, mammary glands from rats fed the HFD during puberty showed a loss of luminal epithelial cell marker gene expression and an increase in mesenchymal cell marker gene expression (Vimentin) and breast cancer invasive genes, similar to human basal-like characteristics, suggesting that a HFD may induce genes associated with a poorer prognosis when consumed during specific times of development (Martinez-Chacin, Keniry, & Dearth, 2014).

4.2 Phytoestrogens

Phytoestrogens are a class of naturally occurring phenolic compounds found in plants or derived from in vivo metabolism of precursors found in plants. The two main subtypes are ligans and isoflavones, which can be found in whole-grain rye bread, red clove, legumes, beans, fruits, vegetables, flaxseed, and soy (Brzezinski & Debi, 1999; Jawaid et al., 2010; Wiseman et al., 2000).

Ligans and isoflavones are shown to be abundant in the blood and urine of populations that are associated with low breast cancer incidence, for example, women of Asian descent (Iwasaki et al., 2008; Wiseman et al., 2000). A diet high in both isoflavones and ligans was shown to associate with later age of menarche, which as we discussed earlier is associated with decreased breast cancer risk (Iwasaki et al., 2008; Mervish et al., 2017). Genistein, found primarily in soy, is the most studied compound in the class of isoflavones. Genistein and other phytoestrogens can act as selective estrogen modulators (SERMs) and can bind to either ERα or ERβ, leading to stimulation or reduction of estrogenic activity and genomic activation (Brzezinski & Debi, 1999; Messina et al., 2013; Rice & Whitehead, 2006).

Asian women have lower breast cancer incidence and mortality compared to NHW women, and this is believed to be the result of a diet with high soy content (Iwasaki et al., 2008; Maskarinec et al., 2010; Messina et al., 2013; Rice & Whitehead, 2006; Wu et al., 2002). The diet appears to be protective in nature during specific times of exposure, especially during pubertal development, as Asian women who immigrated to the United States before onset of pubertal development had higher breast cancer rates, comparable to those observed in NHW women (Iwasaki et al., 2008; Maskarinec et al., 2010; Rice & Whitehead, 2006; Wiseman et al., 2000). Increased soy intake was shown to associate with increases in both PR and ER expression and decreased HER2 expression, both of which are associated with better outcome in women with breast cancer (Messina et al., 2013).

Treatment of estrogen dependent and independent breast cancer cells in vitro with genistein conferred a growth inhibitory effect when administered at high concentrations, but growth stimulatory when administered at lower concentrations (Anderson et al., 2014; Shao, 1998; Yu et al., 2003). When rat models were exposed to genistein through diet (25 and 250 mg), during early stages of development, a reduction of DMBA-induced mammary cancer was observed (20% and 50%, respectively). This was confirmed in a second study that showed a 52% reduction in mammary tumor incidence in mice fed genistein (Anderson et al., 2014; Lamartiniere, Murril, et al., 1998; Lamartiniere, Zhang, et al., 1998). The protective effects observed may have been a result of accelerated mammary cell differentiation during development, as rat models given injections of genistein during neonatal and prepubertal timepoints showed increased proliferation and number of TEBs during pubertal development (21 days of age), but decreased proliferation and number of TEBs at time of carcinogenic treatment (50 days of age) compared to control rats with no genistein exposure (Lamartiniere, Murril, et al., 1998; Lamartiniere, Zhang, et al., 1998). Phytoestrogens may also play a protective role acting as an antioxidant, inhibiting specific enzymes such as tyrosine kinase, topoisomerase II, as well as decreasing angiogenesis, the EGF signaling pathway and local synthesis of estrogen in breast tissue (Lamartiniere, Murril, et al., 1998; Lamartiniere, Zhang, et al., 1998; Rice & Whitehead, 2006; Shao, 1998).

4.3 N-3 Polyunsaturated fatty acids

The three most common N-3 polyunsaturated fatty acids (PUFA) are α-linolenic acid (ALA), commonly found in plant oils such as walnut oil and flaxseed oil and eicosapentaenoic acid (EPA), and docosahexaenoic acid (DHA), both commonly found in fish oils. Consumption of N-3 PUFAs are important for normal metabolism and should be consumed by mammals for a healthy lifestyle. Early exposure to a diet of N-3 PUFAs in mice resulted in a delay in pubertal onset and reduced TEB presence in early puberty, possibly through induced epithelial differentiation (Abdelmagid et al., 2016; Anderson et al., 2014; Leslie et al., 2014; Olivo & Hilakivi-Clarke, 2005). Increased ductal coverage and elongation was also observed in mice fed a N-3 PUFA diet (Abdelmagid et al., 2016; Anderson et al., 2014). As discussed earlier delayed pubertal onset is associated with a decreased in breast cancer risk in women (Abdelmagid et al., 2016). When a HFD with increased N-3 PUFA presence was given to mice, increased proliferation and decreased apoptosis of TEBs occurred; however, with a LFD, the opposite was observed among the TEBs, indicating that dietary effects on pubertal mammary gland development are not linear (Olivo & Hilakivi-Clarke, 2005). The Western diet is also commonly associated with a high N-6 PUFA to N-3 PUFA ratio, averaging at 15:1–16.7:1, compared to an optimal ratio proposed to be 4:1 or lower (Simopoulos, 2002). Early exposure of N-3 PUFAs leads to regulation of pathways involved in energy metabolism, adipose tissue function, inflammation and arachidonic acid specific pathways via reduction of the N-6 PUFA in the mammary gland (Leslie et al., 2014; Liu & Ma, 2014; Luijten et al., 2013; MacLennan et al., 2013; Manni et al., 2011; Olivo & Hilakivi-Clarke, 2005; Zhu et al., 2011).

Americans on average consume a lower amount of N-3 PUFAs compared to Japanese, Norwegian, and Mediterranean populations. Interestingly, populations that consume more N-3 PUFAs have a reduced breast cancer risk (Goodstine et al., 2003; Saadatian-Elahi et al., 2004; Willett, 1999).

A diet with a higher ratio of N-3 PUFAs has been shown to have a protective role against breast cancer through anti-inflammatory effects, modulation of oncogenic protein signaling possibly via disruption of lipid rafts in cell membranes, and increased expression of proteins of cell cycle control and DNA repair (Abdelmagid et al., 2016; Anderson et al., 2014; Bagga et al., 2002; Medvedovic et al., 2009; Saadatian-Elahi et al., 2004; Simopoulos, 2002; Weisburger, 1997). A study in the United States showed that consumption of a diet with increased N-3 PUFAs was associated with a 41% reduction in breast cancer risk in premenopausal and an 11% reduction in postmenopausal women (Goodstine et al., 2003). Studies in mice revealed that N-3 PUFA led to suppressive effects of tumorigenesis via decreased inflammation, cellular proliferation and mammographic density, and increased apoptosis possibly via the inhibition of the MEK/ERK/BAD signaling pathway (Abdelmagid et al., 2016; Anderson et al., 2014; Diorio & Dumas, 2014; Leslie et al., 2014; Liu & Ma, 2014; MacLennan et al., 2013; Manni et al., 2011; Medvedovic et al., 2009; Sun et al., 2011; Zhu et al., 2011). A diet high in N-3 PUFA may also inhibit the effects of an oncogenic N-6 PUFA diet (discussed below) on breast epithelial cells in culture and in the mammary gland itself, as a reduction of arachidonic acid was observed in the mammary gland of mice fed a high N-3 PUFA diet (Bagga et al., 2002; Manni et al., 2011).

4.4 N-6 Polyunsaturated fatty acids

The most common N-6 PUFAs in modern diets originate from vegetable oils, most commonly linoleic acid (LA), which is further broken down in the body to arachidonic acid. Other sources of N-6 PUFAs are eggs, poultry, flaxseed oil, whole-grain breads and cereals. Consumption of N-6 PUFAs are important for normal metabolism and must be consumed by mammals in a specific ratio with N-3 PUFAs for a healthy lifestyle. However, diets in Western developed countries consume much higher levels of N-6 PUFA when compared to N-3 PUFA thereby promoting the pathological effects, leading to the development of diseases such as cardiovascular disease, autoimmune disease, and cancer (Abdelmagid et al., 2016; Simopoulos, 2002). Mice fed a diet high in N-6 PUFAs resulted in earlier pubertal onset, increased TEB presence in early puberty, and increased arachidonic acid in the mammary gland (Abdelmagid et al., 2016; Anderson et al., 2014).

People from Western developed countries consume a diet high in N-6 PUFAs compared to people from Asian countries, where more protective N-3 PUFAs are consumed. Furthermore, as discussed earlier, women from

Western developed countries have higher rates of breast cancer than women from Asian countries, suggesting a cancer-promoting role of N-6 PUFAs (Abdelmagid et al., 2016; Anderson et al., 2014; Goodstine et al., 2003). This could be, in part, due to the increased consumption of processed foods and vegetable oils, leading to increased arachidonic acid in the body, the precursor of leukotrienes and prostaglandin, which are known to be active in carcinogenesis, such as inflammation and cellular proliferation (Abdelmagid et al., 2016; Bagga et al., 2002; Saadatian-Elahi et al., 2004). These reactive compounds can lead to inhibition of gap junctions which alters cellular communication, leading to either apoptosis or the immortalization of the cells, both of which are associated with tumorigenesis (Saadatian-Elahi et al., 2004). Increased ratio of N-6 PUFA to N-3 PUFA has shown to be an important determinant of breast cancer risk, rather than total N-6 PUFA consumption (Bagga et al., 2002). An unbalanced ratio leads to chronic inflammation, modulation of oncogenic protein signaling, and increased proliferation (Abdelmagid et al., 2016; Escrich et al., 2004; MacLennan et al., 2013; Manni et al., 2011).

4.5 Western diet

The Western diet is high in sugar, protein and fat, but low in fruit, grains and vegetables (Uribarri et al., 2010). This diet is also associated with increased consumption of manufactured or processed foods, which are commonly lower in cost and therefore are more widely consumed in low-income populations (Uribarri et al., 2010). Studies have shown that increased consumption of animal protein, caffeinated drinks, and artificially sweetened drinks increases the risk for early onset menarche (Carwile et al., 2015; Mueller et al., 2015; Villamor & Jansen, 2016). On the contra, a diet with increased vegetable protein was shown to associate with a decreased risk for early onset menarche (Villamor & Jansen, 2016). In mice fed a Western diet during puberty, increased number and proliferation of TEBs was observed together with increased mammary ducts (Khan et al., 1994; Xue et al., 1996, 1999).

Asian populations consume as much as 25 g of soy protein or 100 mg of isoflavones a day, this is compared to a Western diet that contains less than 1 g of soy protein or 1 g of isoflavones a day (Fritz et al., 2013). As discussed earlier, women from Asian countries have a lower incidence of breast cancer. Additionally, Asians who immigrate to the United States and adjust to the Western diet lose their protection against development of breast cancer within the first generation (Iwasaki et al., 2008; Maskarinec et al., 2010;

Rice & Whitehead, 2006; Wiseman et al., 2000). The Western diet also consists of increased meat consumption and decreased consumption of fruits/vegetables. Cooked meat contains reactive compounds such as heterocyclic amines and polycyclic aromatic hydrocarbons (Snyderwine, 1998). The most studied of these reactive compounds is 2-amino-1-methyl-6-phenylimidazo[4,5-*b*]pyridine (PhIP). Rats exposed to PhIP lead to tumors in the prostate, colon and mammary gland. Furthermore, human mammary epithelial cells (MCF10A) treated with PhIP underwent genomic and cellular changes, including increased p53 and cyclin dependent kinase inhibitor p21WAF1/CIP1, as well as manipulation of cell cycle and survival as cells were shown to be increasing in the G1 phase (Gooderham et al., 2002). These changes are critical as it can effect genome repair, the removal of damaged cells and lead to mutation acceptance (Gooderham et al., 2002). PhIP was also shown to induce breast cancer invasion and lead to an associated increase in the expression of cathepsin D, cyclooxygenase 2, and matrix metalloproteinase activity (Lauber & Gooderham, 2011).

4.6 Obesity

A sedentary lifestyle, excessive drinking, and poor diet can lead to chronic obesity leading to changes in mammary gland development as well as an increased predisposition to breast cancer (Sundaram et al., 2013). Childhood obesity, as well as a sedentary lifestyle were shown to be associated with an increased risk for early onset menarche (Vani et al., 2013; Villamor & Jansen, 2016). Mice fed a HFD to induce obesity were shown to have enlarged mammary glands with greater adiposity and a less dense ductal tree with decreased branching. The ductal tree was also shown to be surrounded by thick collagen layers and an incomplete myoepithelial layer (Kamikawa et al., 2009).

Chronic obesity has also been shown to lead to increased inflammation leading to formation of crown-like structures (CLS) in mammary glands, structures associated with increased proinflammatory factors (TNFα, IL-1B, COX2, etc.) and aromatase (Bhardwaj et al., 2015; Subbaramaiah et al., 2011). Obesity and CLS also showed to lead to increased NF-κB activation, hypertrophy of adipocytes and increased hormone receptors on tumors (Bhardwaj et al., 2015; Lim et al., 2015a, 2015b; Subbaramaiah et al., 2011). In obese postmenopausal women there was a 50% greater risk of breast cancer development compared to those that were non-obese, potentially from increased adipose tissue and adipokines, which may affect cancer progression site (Gu et al., 2011). Interestingly, caloric restriction in an obesity mouse model was shown to reverse some of the changes caused by obesity; however, weight management alone may be insufficient to completely reverse epigenetic reprogramming and inflammatory signals in the microenvironment of the mammary gland (Bhardwaj et al., 2013, 2015; Rossi et al., 2016).

5. Pubertal intervention strategies

The pubertal window of development, while the hardest to introduce intervention, may dramatically decrease cancer incidence, making it one of the most optimal windows of risk in breast development to target (Martinson et al., 2013). This prompted us to focus our review on the dietary and lifestyle risks of pubertal development and breast cancer. Many pubertal intervention strategies are lifestyle-change based, as pharmacological intervention may be seen as unethical or lead to unknown side effects. Proper education of healthy lifestyle choices during pubertal development is key to introduce the nutritional and lifestyle interventions needed to reduce breast cancer risk in women. Nutritional intervention, such as increased consumption of healthier items associated with decreased breast cancer risk, as discussed above, including vegetables, walnut oil, flaxseed oil, fish, etc., along with decreased consumption of foods associated with increased risk, may help to decrease breast cancer risk overall. Additional lifestyle interventions such as increased physical activity and caloric restriction may also help to decrease adult breast cancer risk.

6. Concluding remarks

The mammary gland and its "lifecycle" of development, specifically pubertal development, is a sensitive time, where cancer susceptibility is high (Biro & Deardorff, 2013; Colditz & Frazier, 1995; Macias & Hinck, 2012; Martinson et al., 2013; Sundaram et al., 2013). External factors, such as diet, can lead to changes in pubertal development ultimately leading to increased breast cancer risk in adulthood. These factors, however, can be mitigated or managed with healthier choices and an active lifestyle, aiding in the reduction of preventable cancers (Dieli-Conwright et al., 2016). Socioeconomic factors demonstrate a link in breast cancer risk such as low-income, poor diet, lack of exercise, and lifestyle choices including smoking and drinking reveal an environmental connection to breast cancer risk and cancer disparity.

References

Abdelmagid, S. A., et al. (2016). Role of n-3 polyunsaturated fatty acids and exercise in breast cancer prevention: Identifying common targets. *Nutrition and Metabolic Insights*, *9*, 71–84.

Ambrosone, C. B., et al. (2014). Associations between estrogen receptor negative breast cancer and timing of reproductive events differ between African-American and European-American women. *Cancer Epidemiology, Biomarkers & Prevention: A Publication of the American Association for Cancer Research, cosponsored by the American Society of Preventive Oncology*, *23*(6), 1115–1120.

Anderson, B. M., et al. (2014). Lifelong exposure to n-3 PUFA affects pubertal mammary gland development. *Applied Physiology, Nutrition, and Metabolism*, *39*(6), 699–706.

Anstey, E. H., Chen, J., Elam-Evans, L., & Perrine, C. G. (2017). Racial and geographic differences in breastfeeding—United States, 2011–2015. *Morbidity and Mortality Weekly Report*, *66*(27), 723–727.

Aupperlee, M. D., et al. (2015). Puberty-specific promotion of mammary tumorigenesis by a high animal fat diet. *Breast Cancer Research: BCR*, *17*, 138.

Bagga, D., et al. (2002). Long-Chain n-3-to-n-6 polyunsaturated fatty acid ratios in breast adipose tissue from women with and without breast cancer. *Nutrition and Cancer*, *42*(2), 180–185.

Ball, S. M. (1998). The development of the terminal end bud in the prepubertal-pubertal mouse mammary gland. *The Anatomical Record*, *250*(4), 459–464.

Bhardwaj, P., et al. (2013). Caloric restriction reverses obesity induced mammary gland inflammation in mice. *Cancer Prevention Research (Philadelphia, Pa.)*, *6*(4), 282–289.

Bhardwaj, P., et al. (2015). Estrogen protects against obesity induced mammary gland inflammation in mice. *Cancer Prevention Research (Philadelphia, Pa.)*, *8*(8), 751–759.

Biro, F. M., & Deardorff, J. (2013). Identifying opportunities for cancer prevention during pre-adolescence and adolescence: Puberty as a window of susceptibility. *The Journal of Adolescent Health: Official Publication of the Society for Adolescent Medicine*, *52*(5 Suppl), S15–S20.

Brady, N. J., Chuntova, P., & Schwertfeger, K. L. (2016). Macrophages: Regulators of the inflammatory microenvironment during mammary gland development and breast cancer. *Mediators of Inflammation*, *2016*, 4549676.

Brzezinski, A., & Debi, A. (1999). Phytoestrogens: The "natural" selective estrogen receptor modulators? *European Journal of Obstetrics, Gynecology, and Reproductive Biology*, *85*(1), 47–51.

Butner, J. D., et al. (2016). A hybrid agent-based model of the developing mammary terminal end bud. *Journal of Theoretical Biology*, *407*, 259–270.

Carwile, J. L., et al. (2015). Sugar-sweetened beverage consumption and age at menarche in a prospective study of US girls. *Human Reproduction (Oxford, England)*, *30*(3), 675–683.

Churpek, J. E., et al. (2015). Inherited predisposition to breast cancer among African American women. *Breast Cancer Research and Treatment*, *149*, 31–39.

Colditz, G. A., & Frazier, A. L. (1995). Models of breast cancer show that risk is set by events of early life: Prevention efforts must shift focus. *Cancer Epidemiology, Biomarkers and Prevention*, *4*(5), 567.

Collaborative Group on Hormonal Factors in Breast Cancer. (2002). Breast cancer and breastfeeding: Collaborative reanalysis of individual data from 47 epidemiological studies in 30 countries, including 50 302 women with breast cancer and 96 973 women without the disease. *The Lancet*, *360*(9328), 187–195.

Cui, Y., et al. (2014). Associations of hormone-related factors with breast cancer risk according to hormone receptor status among white and African-American women. *Clinical Breast Cancer*, *14*(6), 417–425.

Deardorff, J., et al. (2014). Socioeconomic status and age at menarche: An examination of multiple indicators in an ethnically diverse cohort. *Annals of Epidemiology, 24*(10), 727–733.

Dieli-Conwright, C. M., Lee, K., & Kiwata, J. L. (2016). Reducing the risk of breast cancer recurrence: An evaluation of the effects and mechanisms of diet and exercise. *Current Breast Cancer Reports, 8*(3), 139–150.

Diorio, C., & Dumas, I. (2014). Relations of omega-3 and omega-6 intake with mammographic breast density. *Cancer Causes & Control, 25*(3), 339–351.

Economic Policy Institute; n.d. Available from: www.stateofworkingamerica.org.

Escrich, E., et al. (2004). Identification of novel differentially expressed genes by the effect of a high-fat n-6 diet in experimental breast cancer. *Molecular Carcinogenesis, 40*(2), 73–78.

Fritz, H., et al. (2013). Soy, red clover, and isoflavones and breast cancer: A systematic review. *PLoS One, 8*(11), e81968.

Gooderham, N. J., et al. (2002). Molecular and genetic toxicology of 2-amino-1-methyl-6-phenylimidazo[4,5-b]pyridine (PhIP). *Mutation Research, Fundamental and Molecular Mechanisms of Mutagenesis, 506–507*, 91–99.

Goodstine, S. L., et al. (2003). Dietary (n-3)/(n-6) fatty acid ratio: Possible relationship to premenopausal but not postmenopausal breast cancer risk in U.S. women. *The Journal of Nutrition, 133*(5), 1409–1414.

Gu, J.-W., et al. (2011). Postmenopausal obesity promotes tumor angiogenesis and breast cancer progression in mice. *Cancer Biology & Therapy, 11*(10), 910–917.

Inman, J. L., et al. (2015). Mammary gland development: Cell fate specification, stem cells and the microenvironment. *Development, 142*(6), 1028.

Isaac, K., Sundar, F. J., Romanos, G. E., & Rahman, I. (2016). E-cigarettes and flavorings induce inflammatory and prosenescence responses in oral epithelial cells and periodontal fibroblasts. *Oncotarget*, 7(47), 77196–77204.

Iwasaki, M., et al. (2008). Plasma isoflavone level and subsequent risk of breast cancer among Japanese women: A nested case-control study from the Japan Public Health Center-based prospective study group. *Journal of Clinical Oncology, 26*(10), 1677–1683.

Jawaid, K., Crane, S. R., Nowers, J. L., Lacey, M., & Whitehead, S. A. (2010). Long-term genistein treatment of MCF-7 cells decreases acetylated histone 3 expression and alters growth responses to mitogens and histone deacetylase inhibitors. *Journal of Steroid Biochemistry and Molecular Biology, 120*(4–5), 164–171.

Kamikawa, A., et al. (2009). Diet-induced obesity disrupts ductal development in the mammary glands of nonpregnant mice. *Developmental Dynamics, 238*(5), 1092–1099.

Khan, N., Yang, K., Newmark, H., Wong, G., Telang, N., Rivlin, R., et al. (1994). Mammary ductal epithelial cell hyperproliferation and hyperplasia induced by a nutritional stress diet containing four components of a western-style diet. *Carcinogenesis, 15*(11), 2645–2648.

Lamartiniere, C. A., Murril, W. B., Manzolillo, P. A., Zhang, J. X., Barnes, S., Zhang, X., et al. (1998). Genistein alters the ontogeny of mammary gland development and protects against chemically-induced mammary cancer in rats. *Proceedings of the Society for Experimental Biology and Medicine, 217*(3), 358–364.

Lamartiniere, C. A., Zhang, J. X., & Cotroneo, M. S. (1998). Genistein studies in rats: Potential for breast cancer prevention and reproductive and developmental toxicity. *The American Journal of Clinical Nutrition, 68*(6), 1400S–1405S.

Lauber, S. N., & Gooderham, N. J. (2011). The cooked meat-derived mammary carcinogen 2-amino-1-methyl-6-phenylimidazo[4,5-b]pyridine promotes invasive behaviour of breast cancer cells. *Toxicology, 279*(1–3), 139–145.

Leslie, M. A., et al. (2014). Mammary tumour development is dose-dependently inhibited by n-3 polyunsaturated fatty acids in the MMTV-neu(ndl)-YD5 transgenic mouse model. *Lipids in Health and Disease, 13*, 96.

Lim, H. Y., et al. (2015a). Obesity, expression of adipocytokines, and macrophage infiltration in canine mammary tumors. *The Veterinary Journal, 203*(3), 326–331.

Lim, H. Y., et al. (2015b). Effects of obesity and obesity-related molecules on canine mammary gland tumors. *Veterinary Pathology, 52*(6), 1045–1051.

Liu, J., & Ma, D. W. L. (2014). The role of n-3 polyunsaturated fatty acids in the prevention and treatment of breast cancer. *Nutrients, 6*(11), 5184–5223.

Luijten, M., et al. (2013). Lasting effects on body weight and mammary gland gene expression in female mice upon early life exposure to n-3 but not n-6 high-fat diets. *PLoS One, 8*(2), e55603.

Macias, H., & Hinck, L. (2012). Mammary gland development. *Wiley Interdisciplinary Reviews: Developmental Biology, 1*(4), 533–557.

MacLennan, M., & Ma, D. W. L. (2010). Role of dietary fatty acids in mammary gland development and breast cancer. *Breast Cancer Research: BCR, 12*(5), 211.

MacLennan, M. B., et al. (2013). Mammary tumor development is directly inhibited by lifelong n-3 polyunsaturated fatty acids. *Journal of Nutritional Biochemistry, 24*(1), 388–395.

Manni, A., Richie, J. P., Jr., Xu, H., Washington, S., Aliaga, C., Cooper, T. K., et al. (2011). Effects of fish oil and Tamoxifen on preneoplastic lesion development and biomarkers of oxidative stress in the early stages of N-methyl-N-nitrosourea-induced rat mammary carcinogenesis. *International Journal of Oncology, 39*(5), 1153–1164.

Martinez-Chacin, R. C., Keniry, M., & Dearth, R. K. (2014). Analysis of high fat diet induced genes during mammary gland development: Identifying role players in poor prognosis of breast cancer. *BMC Research Notes, 7*, 543.

Martinson, H. A., et al. (2013). Developmental windows of breast cancer risk provide opportunities for targeted chemoprevention. *Experimental Cell Research, 319*(11), 1671–1678.

Maskarinec, G., Verheus, M., & Tice, J. A. (2010). Epidemiologic studies of isoflavones mammographic density. *Nutrients, 2*(1), 35–48.

Medvedovic, M., et al. (2009). Influence of fatty acid diets on gene expression in rat mammary epithelial cells. *Physiological Genomics, 38*(1), 80–88.

Mervish, N. A., et al. (2017). Peripubertal dietary flavonol and lignan intake and age at menarche in a longitudinal cohort of girls. *Pediatric Research, 82*(2), 201–208.

Messina, M., Caan, B. J., Abrams, D. I., Hardy, M. A., & Maskarinec, G. (2013). It's time for clinicians to reconsider their proscription against the use of soyfoods by breast cancer patients. *Oncology, 27*(5), 430–437.

Mueller, N. T., et al. (2015). Consumption of caffeinated and artificially sweetened soft drinks is associated with risk of early menarche. *The American Journal of Clinical Nutrition, 102*(3), 648–654.

Okasha, M., et al. (2003). Exposures in childhood, adolescence and early adulthood and breast cancer risk: A systematic review of the literature. *Breast Cancer Research and Treatment, 78*(2), 223–276.

Olivo, S. E., & Hilakivi-Clarke, L. (2005). Opposing effects of prepubertal low- and high-fat n-3 polyunsaturated fatty acid diets on rat mammary tumorigenesis. *Carcinogenesis, 26*(9), 1563–1572.

Olson, L. K., et al. (2010). Pubertal exposure to high fat diet causes mouse strain-dependent alterations in mammary gland development and estrogen responsiveness. *International Journal of Obesity (2005), 34*(9), 1415–1426.

Paine, I. S., & Lewis, M. T. (2017). The terminal end bud: The little engine that could. *Journal of Mammary Gland Biology and Neoplasia, 22*, 1–16.

Phipps, A. I., & Li, C. I. (2014). Breastfeeding and triple-negative breast cancer: Potential implications for racial/ethnic disparities. *JNCI: Journal of the National Cancer Institute, 106*(10), dju281.

Ramnitz, M. S., & Lodish, M. B. (2013). Racial disparities in pubertal development. *Seminars in Reproductive Medicine, 31*(5), 333–339.

Reagan, P. B., et al. (2012). African-American/white differences in the age of menarche: Accounting for the difference. *Social Science & Medicine (1982)*, *75*(7), 1263–1270.

Rice, S., & Whitehead, S. A. (2006). Phytoestrogens and breast cancer—Promoters or protectors? *Endocrine-Related Cancer*, *13*(4), 995–1015.

Rossi, E. L., et al. (2016). Obesity-associated alterations in inflammation, epigenetics, and mammary tumor growth persist in formerly obese mice. *Cancer Prevention Research*, *9*(5), 339.

Saadatian-Elahi, M., et al. (2004). Biomarkers of dietary fatty acid intake and the risk of breast cancer: A meta-analysis. *International Journal of Cancer*, *111*(4), 584–591.

Shao, Z. M. Z. (1998). Genistein exerts multiple suppressive effects on human breast carcinoma cells. *Cancer Research*, *58*(21), 4851–4857.

Shrivastava, S., Shrivastava, P., & Ramasamy, J. (2016). Exploring the role of dietary factors in the development of breast cancer. *Journal of Cancer Research and Therapeutics*, *12*(2), 493–497.

Simopoulos, A. P. (2002). The importance of the ratio of omega-6/omega-3 essential fatty acids. *Biomedicine & Pharmacotherapy*, *56*(8), 365–379.

Smith, P. J., et al. (2005). Statistical methodology of the National Immunization Survey, 1994–2002. *Vital Health Statistics 2*, *138*, 1–55.

Snyderwine, E. G. (1998). Diet and mammary gland carcinogenesis. *Recent Results in Cancer Research*, *152*, 3–10.

Snyderwine, E. G., Davis, C. D., Schut, H. A., & Roberts-Thomson, S. J. (1998). Proliferation, development and DNA adduct levels in the mammary gland of rats given 2-amino-1-methyl-6-phenylimidazo[4,5-b]pyridine and a high fat diet. *Carcinogenesis*, *19*(7), 1209–1215.

Snyderwine, E. G., et al. (1998). Mammary gland carcinogenicity of 2-amino-l-methyl-6-phenylimidazo[4,5-b]pyridine in Sprague-Dawley rats on high- and low-fat diets. *Nutrition and Cancer*, *31*(3), 160–167.

Song, F., et al. (2014). RAGE regulates the metabolic and inflammatory response to high-fat feeding in mice. *Diabetes*, *63*(6), 1948–1965.

Sternlicht, M. D. (2006). Key stages in mammary gland development: The cues that regulate ductal branching morphogenesis. *Breast Cancer Research*, *8*(1), 201.

Sternlicht, M. D., et al. (2006). Hormonal and local control of mammary branching morphogenesis. *Differentiation; Research in Biological Diversity*, *74*(7), 365–381.

Subbaramaiah, K., et al. (2011). Obesity is associated with inflammation and elevated aromatase expression in the mouse mammary gland. *Cancer Prevention Research (Philadelphia, Pa.)*, *4*(3), 329–346.

Sun, H., et al. (2011). Omega-3 fatty acids induce apoptosis in human breast cancer cells and mouse mammary tissue through syndecan-1 inhibition of the MEK-Erk pathway. *Carcinogenesis*, *32*(10), 1518–1524.

Sundaram, S., Johnson, A. R., & Makowski, L. (2013). Obesity, metabolism and the microenvironment: Links to cancer. *Journal of Carcinogenesis*, *12*, 19.

Uribarri, J., et al. (2010). Advanced glycation end products in foods and a practical guide to their reduction in the diet. *Journal of the American Dietetic Association*, *110*(6), 911-916.e12.

U.S. Department of Health and Human Service; n.d. Available from: www.minorityhealth.hhs.gov.

Vani, K. R., et al. (2013). Menstrual abnormalities in school going girls—Are they related to dietary and exercise pattern? *Journal of Clinical and Diagnostic Research: JCDR*, 7(11), 2537–2540.

Villamor, E., & Jansen, E. C. (2016). Nutritional determinants of the timing of puberty. *Annual Review of Public Health*, *37*, 33–46.

Warner, E. T., et al. (2013). Estrogen receptor positive tumors: Do reproductive factors explain differences in incidence between black and white women? *Cancer Causes & Control*, *24*(4), 731–739.

Weisburger, J. H. (1997). Dietary fat and risk of chronic disease: Mechanistic insights from experimental studies. *Journal of the American Dietetic Association*, *97*(7), S16–S23.

Willett, W. C. (1999). Goals for nutrition in the year 2000. *CA: A Cancer Journal for Clinicians*, *49*(6), 331–352.

Wiseman, H., et al. (2000). Isoflavone phytoestrogens consumed in soy decrease F(2)-isoprostane concentrations and increase resistance of low-density lipoprotein to oxidation in humans. *American Journal of Clinical Nutrition*, *72*(2), 395–400.

World Health Organization. (2017). *Breastfeeding*. Available from: http://www.who.int/topics/breastfeeding/en/.

Wu, A. H., et al. (2002). Adolescent and adult soy intake and risk of breast cancer in Asian-Americans. *Carcinogenesis*, *23*(9), 1491–1496.

Xue, L., et al. (1996). Model of mouse mammary gland hyperproliferation and hyperplasia induced by a western-style diet. *Nutrition and Cancer*, *26*(3), 281–287.

Xue, L., et al. (1999). Influence of dietary calcium and vitamin D on diet-induced epithelial cell hyperproliferation in mice. *JNCI: Journal of the National Cancer Institute*, *91*(2), 176–181.

Yu, Z., Zhang, L., & Wu, D. (2003). Genistein induced apoptosis in MCF-7 and T47D cells. *Wei Sheng Yan Jiu = Journal of Hygiene Research*, *32*(2), 125–127.

Zhao, Y., et al. (2013). Pubertal high fat diet: Effects on mammary cancer development. *Breast Cancer Research: BCR*, *15*(5), R100.

Zhu, Z., et al. (2011). Mammary gland density predicts the cancer inhibitory activity of the N-3 to N-6 ratio of dietary fat. *Cancer Prevention Research*, *4*(10), 1675.

Zhu, Y., et al. (2016). Pubertal and adult windows of susceptibility to a high animal fat diet in Trp53-null mammary tumorigenesis. *Oncotarget*, 7(50), 83409–83423.

CHAPTER FOUR

BMI, physical activity, and breast cancer subtype in white, black, and Sea Island breast cancer survivors

Marvella E. Ford[a,b,c,*], Colleen E. Bauza[a], Victoria J. Findlay[d,e], David P. Turner[d,e], Latecia M. Abraham[f], Leslie A. Moore[g], Gayenell Magwood[d,h], Anthony J. Alberg[i], Kadeidre Gaymon[h], Kendrea D. Knight[a], Ebony Hilton[j], Angela M. Malek[a], Rita M. Kramer[k], Lindsay L. Peterson[l], Mathew J. Gregoski[m], Susan Bolick[n], Deborah Hurley[n], Catishia Mosley[n], Tonya R. Hazelton[h], Dana R. Burshell[o], Lourdes Nogueira[e], Franshawn Mack[p], Erika T. Brown[q], Judith D. Salley[p], Keith E. Whitfield[r], Nestor F. Esnaola[s,t,u,v], Joan E. Cunningham[w]

[a]Professor, Department of Public Health Sciences, Medical University of South Carolina, Charleston, SC, United States
[b]Associate Director, Population Sciences and Cancer Disparities, Hollings Cancer Center, Medical University of South Carolina, Charleston, SC United States
[c]SmartState Endowed Chair in Cancer Disparities Research, South Carolina State University, Orangeburg, SC, United States
[d]Hollings Cancer Center, Medical University of South Carolina, Charleston, SC, United States
[e]Department of Pathology and Laboratory Medicine, Medical University of South Carolina, Charleston, SC, United States
[f]Department of Library Science and Informatics, Medical University of South Carolina, Charleston, SC, United States
[g]College of Medicine, Medical University of South Carolina, Charleston, SC, United States
[h]College of Nursing, Medical University of South Carolina, Charleston, SC, United States
[i]Department of Epidemiology and Biostatistics, University of South Carolina, Columbia, SC, United States
[j]Department of Anesthesiology and Perioperative Medicine, Medical University of South Carolina, Charleston, SC, United States
[k]Department of Hematology/Oncology, Medical University of South Carolina, Charleston, SC, United States
[l]Department of Medicine, Washington University in St. Louis, St. Louis, MO, United States
[m]Department of Arts & Sciences, Campbell University, Buies Creek, NC, United States
[n]South Carolina Department of Health and Environmental Control, Columbia, SC, United States
[o]South Carolina Clinical & Translational Research Institute, Clinical and Translational Science Award, Medical University of South Carolina, Charleston, SC, United States
[p]Chair, Department of Biological & Physical Sciences, South Carolina State University, Orangeburg, SC, United States
[q]Morehouse School of Medicine, Atlanta, GA, United States
[r]Provost and Senior Vice President for Academic Affairs, Wayne State University, Detroit, MI, United States
[s]Professor of Surgery, Houston Methodist Academic Institute/Weill Cornell Medical College, Houston, TX, United States
[t]Division Chief of Surgical Oncology and Gastrointestinal Surgery, Department of Surgery, Houston Methodist Hospital, Houston, TX, United States
[u]Surgical Director and Associate Director of Community Engagement and Cancer Control, Houston Methodist Cancer Center, Houston, TX, United States
[v]Katz Investigator, Houston Methodist Research Institute, Houston, TX, United States

Advances in Cancer Research, Volume 146
ISSN 0065-230X
https://doi.org/10.1016/bs.acr.2020.01.005

[w]The National Coalition of Independent Scholars, San Antonio, TX, United States
*Corresponding author: e-mail address: fordmar@musc.edu

Contents

Abstract

Higher BMI, lower rates of physical activity (PA), and hormone receptor-negative breast cancer (BC) subtype are associated with poorer BC treatment outcomes. We evaluated the prevalence of high BMI, low PA level, and BC subtype among survivors with white/European American (EA) and African American (AA) ancestry, as well as a distinct subset of AAs with Sea Island/Gullah ancestry (SI). We used the South Carolina Central Cancer Registry to identify 137 (42 EAs, 66 AAs, and 29 SIs) women diagnosed with BC and who were within 6–21 months of diagnosis. We employed linear and logistic regression to investigate associations between BMI, PA, and age at diagnosis by racial/ethnic group. Most participants (82%) were overweight/obese ($P = 0.46$). BMI was highest in younger AAs ($P = 0.02$). CDC PA guidelines (≥150 min/week) were met by only 28% of participants. The frequency of estrogen receptor (ER)-negative BC subtype was lower in EAs and SIs than in AAs ($P < 0.05$). This is the first study to identify differences in obesity and PA rates, and BC subtype in EAs, AAs, and SIs. BMI was higher, PA rates were lower, and frequency of ER-negative BC was higher in AAs as compared to EAs and SIs. This study highlights the need to promote lifestyle interventions among BC survivors, with the goal of reducing the likelihood of a BC recurrence. Integrating dietary and PA interventions into ongoing survivorship care is essential. Future research could evaluate potential differential immune responses linked to the frequency of triple negative BC in AAs.

1. Introduction

Breast cancer (BC) is the second leading cause of cancer death among women in the United States (US Cancer Statistics Working Group, 2017). Compared with European American (EA)/white women, African American/Black (AA) women have disproportionately poorer survival after a BC diagnosis, resulting in 38% higher BC mortality rates for 2009–2013 (American Cancer Society, 2013; Campbell et al., 2012).

Obesity has also been linked to poorer BC outcomes, with overweight or obese women (body mass index [BMI] ≥25) having a significantly higher risk of BC recurrence than women with normal BMI levels (Demark-Wahnefried, Campbell, & Hayes, 2012; Fabian, 2012; Kamineni et al., 2013). United States' obesity (BMI ≥30) has reached epidemic proportions, with an age-adjusted prevalence of 36% among EA women and 57% among AA women in 2011–2014 (Ogden, Carroll, Fryar, & Flegal, 2015). In the United States, AAs experience the highest rates of obesity, providing one possible reason for their poorer survival (American Cancer Society, 2014).

Another potential contributor to the racial/ethnic disparity in BC survival is physical activity (Thompson, Owusu, Nock, Li, & Berger, 2014). Women with BC who are physically active have improved survival rates compared to those who are not (Thompson et al., 2014). However, treatment of early stage BC often results in fatigue, functional decline and weight gain which makes increasing the amount of physical activity more difficult. Compounding the challenges faced by women with a BC diagnosis, many obese, sedentary survivors have co-morbid illnesses such as diabetes, hyperlipidemia and hypertension, all of which are likely to improve with increased physical activity.

The purpose of the present study was to compare the prevalence of high BMI and level of physical activity by racial/ethnic group among BC survivors in South Carolina. A unique feature of this study is the ability to examine the frequency of the triple negative BC (TNBC) subtype, as it is well known that TNBC occurs with greater frequency in AA than in EA women.

Also novel is the inclusion not only of EA and AA but also Sea Islanders/Gullah (SI) women, a distinct subset of AAs that is culturally and genetically distinct (Tishkoff et al., 2009; Zakharia et al., 2009). The SIs are direct descendants of Africans from the west coast of Africa. Geographic isolation has allowed the SIs, more than any other AA group, to preserve elements of their African cultural heritage and to retain one of the lowest rates of genetic admixture of any AA group in the United States (less than 4%, compared to an average 15% in the United States AA population) (McLean et al., 2005).

The presence of the SI population in South Carolina provides a unique opportunity to explore potential biologic differences that contribute to disparities in cancer morbidity and mortality in AAs in South Carolina and beyond (Tishkoff et al., 2009; Zakharia et al., 2009), and to investigate whether, among survivors of BC, overweight/obesity and physical activity differ among AA subgroups. The present study provides the first known such comparisons across these three racial/ethnic groups: EAs, AAs without SI ancestry (non-SI AAs), and SIs.

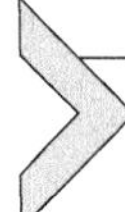

2. Methods

2.1 Institutional Review Board (IRB) protection

We obtained approval to conduct the study from the IRB at the Medical University of South Carolina (MUSC) and from the South Carolina Central Cancer Registry (SCCCR) of the South Carolina Department of Health and Environmental Control (South Carolina DHEC) (South Carolina Department of Health and Environmental Control, 2017). Consent was obtained from each study participant.

2.2 Breast cancer (BC) case ascertainment

Case ascertainment and data collection steps are illustrated in the CONSORT diagram (Fig. 1). Potential participants consisted of adult women with a diagnosis of invasive BC of known grade at age ≥21 years, and whose race was recorded in the SCCCR as either EA or AA. Cases were ascertained between May 2012 and October 2013 by SCCCR staff within 6–21 months (mean 12 months) of diagnosis, and then interviewed by study personnel within 6 months of SCCCR staff referral. All interviews were completed by December 2013.

After receiving passive physician approval for each potentially eligible case identified, SCCCR staff contacted the participants by telephone and referred those who declared an interest in participating ("opted in") to study personnel. The target sample size was 30 participants in each of the three racial/ethnic groups.

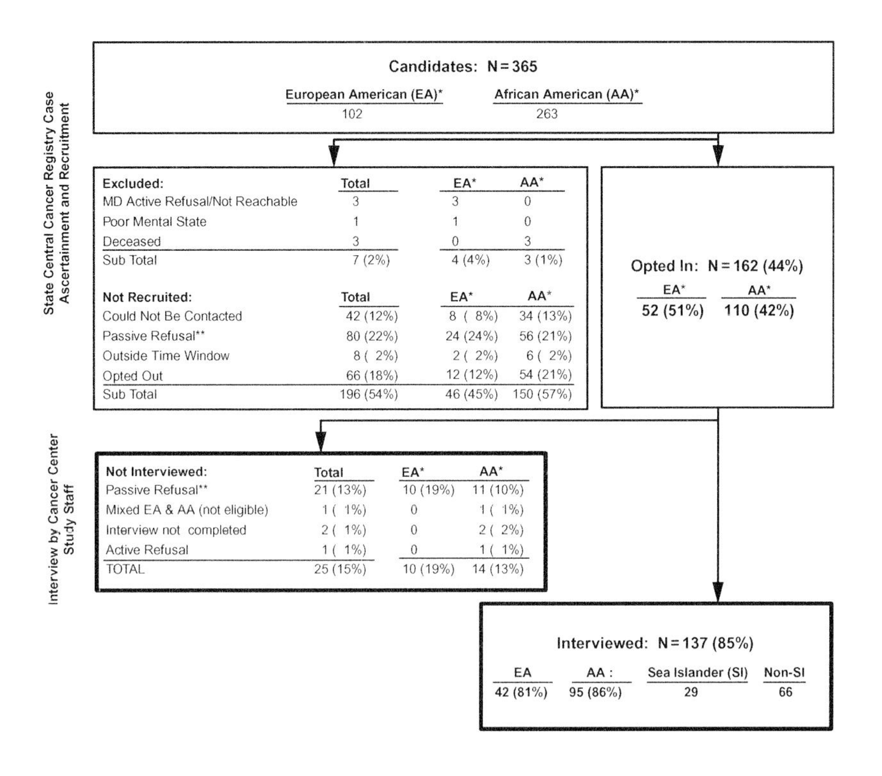

Fig. 1 CONSORT diagram.

2.3 Data collection and ethnicity definition

The SCCCR staff provided data on BC characteristics (date of diagnosis, stage, tumor markers) for each participating case. Trained study interviewers contacted each woman referred by the SCCCR staff and administered a telephone interview. Ethnicity of each participant was categorized on the basis of the ethnicity of parents as self-reported during the interview and according to a previously developed algorithm to confirm SI ancestry: EA if the participant considered both parents to be of European origin, or AA if both parents were AA. Among AAs, participants were sub-classified as SIs if both parents were born in regions of South Carolina classified as SI geographic regions (i.e., ≤ 30 miles from the Atlantic coast) or AAs without SI ancestry (non-SI AAs).

AA participants whose parental racial/ethnic ancestry was unknown were not included in the present analysis. Participants who reported Asian or Hispanic ancestry or who were of mixed race were also excluded from this analysis.

Other self-reported data collected during the telephone interview included age, education, height, weight, primary leisure time physical activity during the past month, and frequency and usual time spent in this activity. Questionnaire items were selected from the National Health Interview Survey and from the CDC Behavioral Risk Factor Surveillance Survey (BRFSS) (Centers for Disease Control and Prevention (CDC), 2011, 2018). BMI (kg/m^2) was calculated from self-reported height and weight as [weight (lbs)/height (in.)^2] (Centers for Disease Control and Prevention (CDC), 2015). Published CDC BMI categories and CDC 2008 physical activity guidelines (Centers for Disease Control and Prevention (CDC), 2015, 2017) were used to characterize BMI and physical activity guideline adherence in the study sample. BMI $<$18.5 is considered underweight, 18.5–24.9 normal weight, 25.0–29.9 overweight, 30.0–34.9 obese, and $\geq$35.0 as extremely obese.

Duration of moderate aerobic activity per week was calculated from the reported duration of primary type of activity or exercise per episode, multiplied by the number of episodes per week during the past month. This was adjusted for intensity according to metabolic equivalents of that activity, where 1 min of vigorous activity counted as 2 min of moderate activity; light or non-aerobic activity or activity for less than 10 min per session was not included. Activity level was then categorized according to the CDC-recommended guidelines of 150 min moderate aerobic activity per week (National Center for Chronic Disease Prevention and Health Promotion. Division of Nutrition Physical Activity and Obesity. CDC, 2013).

To assess how representative our study population was of statewide incident invasive BC cases, we compared age at diagnosis, prevalence of late stage (regional or distant) and prevalence of triple negative disease to data for all invasive cases of known stage diagnosed during 2011–2012, among adult EA and AA women (provided by the SCCCR). We also compared educational attainment, BMI and physical activity among participants with 2012–2013 South Carolina BRFSS data to EA and AA adult women of the same age range as participants; these data represent the general female population and are not restricted to those with invasive BC.

2.4 Evaluation of triple negative breast cancer (TNBC) in a sub-sample of the three population groups

While it is well known that TNBC occurs with greater frequency in AA than in EA women, what is unknown is the frequency of TNBC in the SI population. Therefore, we recruited and collected saliva samples from a sub-sample of 86 participants from the three population groups (approximately 30 participants per group) who had been diagnosed with BC in the past 6–21 months.

2.5 Statistical analysis

Participant demographic and tumor characteristics, BMI, and physical activity were summarized using standard descriptive statistics. In accordance with the primary purpose of this study, statistical analyses examined differences among and between the three racial/ethnic groups: EAs, non-SI AAs, and SIs. Differences were evaluated statistically using Student's t-test, chi-square, Fisher exact test, Wilcoxon rank-sum, and Kruskal-Wallis H non-parametric analysis, as appropriate. Simple linear and logistic regression modeling techniques were used to investigate associations between BMI, physical activity, and age at diagnosis. Statistical analyses were conducted using STATA 10.0 (StataCorp, 2007). Tests were two-tailed, except where testing associations between physical activity and BMI, which were one-tailed (based upon the hypothesis that greater physical activity would be associated with reduced BMI); P-values <0.05 were considered statistically significant.

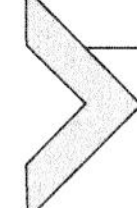

3. Results

3.1 Study population

A total of 365 potentially eligible participants were ascertained by SCCCR staff, of whom 162 (44%) opted into the study and were referred to the study's data collection staff. Among these, 137 (85%) met all eligibility criteria and were interviewed: 42 EAs, 66 non-SI AAs, and 29 SIs (CONSORT diagram: Fig. 1). All BC cases had been diagnosed during 2012 and 2013.

3.2 Sociodemographic characteristics and BC stage

Age at diagnosis ranged from 37.9 to 89.4 years, with differences in mean age significant only between non-SI AAs and EAs ($P=0.044$) (Table 1). Educational attainment did not differ statistically between AAs and EAs

Table 1 Participant and breast cancer characteristics in the three racial/ethnic groups.

	EA		AA				P-value[a]			
	($N = 42$)		Non-SI ($N = 66$)		SI ($N = 29$)					
	N	(%)	N	(%)	N	(%)	Three groups	EA vs non-SI AA	Non-SI AA vs SI	SI vs EA
Age at diagnosis (years)										
Mean (std dev)	61.6	(12.9)	56.7	(12.0)	60.2	(10.4)	0.094	**0.044**	0.152	0.609
Median (range)	62.1	(37.9–86.6)	54.9	(37.1–89.4)	57.4	(45.2–81.4)				
Education										
<HS diploma	3	(7)	12	(18)	8	(28)	0.190	0.562	0.292	**0.029**
HS diploma/12 years	12	(29)	18	(27)	8	(28)				
Trade/some college	16	(38)	19	(29)	3	(10)				
College degree	6	(14)	10	(15)	7	(24)				
Post graduate degree	5	(12)	7	(11)	3	(10)				
<HS or HS diploma	15	(36)	30	(45)	16	(55)	0.118	0.522	0.145	**0.029**
Trade/some college	16	(38)	19	(29)	3	(10)				
College degree (+)	11	(26)	17	(26)	10	(34)				
Breast cancer										
Localized	33	(79)	31	(47)	21	(72)	**0.011**	**0.004**	0.074	0.886
Regional	8	(19)	28	(42)	7	(24)				
Distant	1	(2)	7	(11)	1	(3)				
Regional/distant	9	(21)	35	(53)	8	(28)	**0.002**	**0.001**	**0.026**	0.582
Triple negative	3/39	(7.7)	14/61	(23.0)	2/21	(9.5)	0.102	0.058	0.220	1.000

[a]Values in bold font type indicate P-value <0.05.
EA, European American; AA, African American; non-SI AA, non-SI African American; SI, Sea Islander; std dev, standard deviation; HS, high school; triple negative: negative for expression of estrogen receptor, progesterone receptor and HER2.

or between AAs and SIs. However, EAs were more likely than SIs to report having a high school diploma or more education ($P = 0.029$).

Racial/ethnic differences were observed in stage at diagnosis (among groups: $P = 0.002$), with late stage disease most common among non-SI AAs (21% of EAs, 53% of non-SI AAs and 28% of SIs). Interestingly, late stage was only half as common among SIs as non-SI AAs ($P = 0.026$).

Among those with complete tumor marker data ($N = 121$), TNBC (negative for each of estrogen receptor [ER], progesterone receptor [PR], and human epidermal growth factor receptor 2 [HER-2] expression) was found in 8% of EAs, 23% of non-SI AAs, and 10% of SIs, with no significant differences among or between groups. As with stage, prevalence of TNBC in SIs was similar to EAs, and about half that seen in non-SI AAs. In interpreting this it should be noted that HER-2 data were not available for 28% of SI cases, compared with only 7%–8% of EA and non-SI AA cases.

3.3 Overweight/obesity

Most participants were overweight/obese (BMI ≥ 25; Table 2), a finding that did not vary significantly among racial/ethnic groups (79% of EAs, 80% of non-SI AAs, and 90% of SIs) or between AA groups. Non-SI AAs and SIs were each more likely than EAs to have extreme obesity (BMI ≥ 35.0: 33% and 28% respectively, vs 19% of EAs), not statistically significant but nevertheless of concern.

3.4 Physical activity

Most participants reported having some physical activity outside of their regular job (81% of EAs, 62% of non-SI AAs, and 79% of SIs; $P = 0.071$; Table 2), although there was not a significant difference among or between groups. Adherence to the CDC-recommended guidelines of at least 150 min per week of moderate aerobic physical activity was low (41% of EAs, 29% of non-SI AAs, and 39% of SIs), with no significant differences among or between groups but a pattern of lowest activity was observed among non-SI AAs in both metrics.

Duration of moderate aerobic activity varied considerably from 0 to 2160 min/week, was highest among EAs (median 120 min/week; 52 min/week among non-SI AAs, and 60 min/week among SIs), but not significantly different among groups ($P = 0.053$) although greater among EAs than non-SI AAs ($P = 0.021$).

Table 2 BMI and physical activity in the three racial/ethnic groups.

	EA		AA				*P*-value[a]			
	(*N* = 42)		Non-SI (*N* = 66)		SI (*N* = 29)					
	N	(%)	*N*	(%)	*N*	(%)	Three groups	EA vs non-SI AA	Non-SI AA vs SI	SI vs EA
Body mass index (BMI), kg/m^2										
Mean (std dev)	29.4	(5.1)	31.9	(7.5)	32.0	(6.1)	0.140	0.051	0.943	0.061
Median (range)	29.5	(20.8–40.2)	31.0	(16.6–48.4)	30.7	(22.0–51.6)				
Underweight/normal (≤24.9)[b]	9	(21)	13	(20)	3	(10)	0.555	0.436	0.490	0.561
Overweight (25.0–29.9)	15	(36)	18	(27)	9	(31)				
Obese (30.0–34.9)	10	(24)	13	(20)	9	(31)				
Extreme obesity (≥35.0)	8	(19)	22	(33)	8	(28)				
Overweight/obese (≥25.0)	33	(79)	53	(80)	26	(90)	0.460	0.812	0.375	0.336
Physical activity										
None	8	(19)	25	(38)	6	(21)	0.071	0.053	0.153	1.000
Some	34	(81)	41	(62)	23	(79)				
Main activity (*N* = 98)										
Walking, hiking, treadmill	25	(74)	30	(73)	17	(74)	1.000	1.000	1.000	1.000

Gym, swim, group exercise	5	(15)	6	(15)	3	(13)				
Gardening, housework, other	4	(12)	5	(12)	3	(13)				
Minutes per week[c,d]										
Mean (std dev)	318	(498)	119	(207)	212	(375)	–	–	–	–
Median (range)	120	(0–2160)	52	(0–1260)	60	(0–1800)	0.053	**0.021**	0.160	0.424
None	9	(22)	26	(39)	6	(21)	**0.009**	**0.002**	0.246	0.249
0–149	15	(37)	21	(32)	11	(39)				
150–299	3	(7)	14	(21)	6	(21)				
300+	14	(34)	5	(8)	5	(19)				
CDC activity guidelines met[e]										
No	24	(59)	47	(71)	17	(61)	0.350	0.177	0.318	0.856
Yes	17	(41)	19	(29)	11	(39)				

[a]Values in bold font type indicate P-value <0.05.
[b]Includes one underweight Non-SI AA participant.
[c]Minutes per week: includes moderate activity, plus vigorous activity where 1 min of vigorous activity counts as 2 min of moderate activity; excludes activity for less than 10 min per session, or of less than 3.0 metabolic equivalents (National Center for Chronic Disease Prevention and Health Promotion. Division of Nutrition Physical Activity and Obesity. CDC, 2013).
[d]Two participants (one EA, one SI AA) did not provide information as to frequency or duration of physical activity and are excluded from this analysis.
[e]CDC guidelines: at least 150 min per week moderate aerobic physical activity, as defined above.
BMI, body mass index; EA, European American; AA, African American; SI, Sea Islander; std dev, standard deviation; CDC, Centers for Disease Control and Prevention.

3.5 Associations with age

BMI and physical activity each were significantly associated with age at interview, consistent in terms of the direction of the association across all racial/ethnic groups except SIs (Table 3). Older participants were less likely to have higher BMIs (reduction of 0.11 BMI unit per increased year of age, $P = 0.024$), particularly non-SI AAs (reduction of 0.19 BMI unit per year, $P = 0.015$). Consistent with this finding, older participants were more likely to have met the CDC physical activity guidelines (6% increased odds per year, $P < 0.001$), particularly EAs (11%, $P = 0.004$) and non-SI AAs (6%, $P = 0.028$).

3.6 Physical activity and BMI

Having met the CDC physical activity guidelines was found to reduce BMI (0.8–2.5 units) within each racial/ethnic group and among all participants combined, but not significantly (P-values 0.086–0.742, Table 3). Reduction was least among SIs.

3.7 Comparison with state-level population-based data

Table 4 shows that our study sample compared well with all cases of invasive BC of known stage diagnosed in South Carolina among adult EA and AA women during 2011–2012 (SCCCR data). Mean ages at diagnosis were very similar: EAs in our sample were on average 1.4 years younger than in state data, and non-SI AAs (comprising the greater part of South Carolina AAs, for whom SI ethnicity data are not available) were 2.0 years younger. Of interest, SCCCR AA data were within the range of data from the study's non-SI AA and SI study sample on all parameters: age, late state and TNBC.

By next comparing our study sample with 2012–2013 South Carolina BRFSS population-based, state-level data for EA and AA women corresponding to the age range of our study sample, we thus compared our cancer cases with the general population. Women in our study were somewhat more likely than the BRFSS population to have attained greater than a high school diploma (EAs: 64% vs 57% of BRFSS; AAs: 55% of non-SI AAs and 45% of SIs, vs 44% of BRFSS). Overweight/obesity was somewhat more prevalent among EA participants (79% vs 61% of BRFSS), but rather similar among AAs (80% of non-SI AAs and 90% of SIs, vs 84% of BRFSS).

Table 3 BMI, physical activity and age in the three racial/ethnic groups.

	Population	*N*	Coef. or OR	*P-value*[a]	(95% CI)
Increased BMI (per unit)			Coef.		
Age at interview (per year)	All[b]	137	−0.11	**0.024**	(−0.20 to −0.15)
	EA	42	−0.07	0.237	(−0.20 to 0.05)
	Non-SI AA	66	−0.19	**0.015**	(−0.35 to −0.04)
	SI	29	+0.07	0.537	(−0.16 to 0.30)
Overweight/obesity vs normal[c]			OR		
Age at interview (per year)	All[b]	137	0.96	**0.020**	(0.92 to 0.99)
	EA	42	0.95	0.091	(0.89 to 1.01)
	Non-SI AA	66	0.94	**0.034**	(0.89 to 1.00)
	SI	29	1.07	0.359	(0.93 to 1.23)
CDC physical activity (PA) guidelines met vs not met[d]			OR		
Age at Interview (per year)	All[b]	135	1.06	**<0.001**	(1.03 to 1.10)
	EA	41	1.11	**0.004**	(1.03 to 1.19)
	Non-SI AA	66	1.06	**0.028**	(1.01 to 1.11)
	SI	28	1.01	0.704	(0.94 to 1.09)
Increased BMI (per unit)[c,f]			Coef.	*P*-value[f]	95% CI[f]
(a) All participants	All[b]	135	−2.11	0.086	(−4.53 to 0.30)
	EA	41	−2.52	0.125	(−5.78 to 0.73)
	Non-SI AA	66	−2.44	0.249	(−6.64 to 1.75)
	SI	28	−0.82	0.742	(−5.87 to 4.23)

Continued

Table 3 BMI, physical activity and age in the three racial/ethnic groups.—Cont'd

	Population	*N*	Coef. or OR	*P-value*[a]	(95% CI)
Overweight/obesity vs normal[c,e,f]			OR	*P*-value[f]	95% CI[f]
(a) All participants	All[b]	135	0.49	0.118	(0.20 to 1.20)
	EA	41	0.26	0.094	(0.05 to 1.25)
	Non-SI AA	66	0.57	0.393	(0.16 to 2.05)
	SI	28	1.33	0.824	(0.11 to 16.74)

[a]Values in bold font type indicate *P*-value <0.05.
[b]Adjusted for race/ethnicity.
[c]BMI category "Normal" here includes one underweight non-SI AA participant.
[d]Two participants (one EA, one SI AA) did not provide information as to frequency or duration of PA and are excluded from the analysis of CDC PA guideline achievement.
[e]Not adjusted for age at diagnosis, to avoid over-adjustment.
[f]*P*-value and 95% confidence intervals are for one-tailed tests of the hypothesis that increased PA is associated with reduced BMI (coefficient < zero) or normal BMI (OR < 1.00).
BMI, body mass index; CI, confidence interval; Coef., regression coefficient; OR, odds ratio; *P*, *P*-value; EA, European America; AA, African American; SI, Sea Islander.

Table 4 Demographics, cancer stage, BMI and physical activity: Comparing study population to state-level data for South Carolina.

	EA		AA		
	Current study	State data, %	Current study		
			Non-SI	SI	
	N = 42, %		*N* = 66, %	*N* = 29, %	State data, *N* = 6688, %
Mean age at diagnosis (years)	61.6	63.0[a]	56.7	60.2	58.7[a]
Breast cancer					
Late stage (regional/distant vs localized)	21	35[a]	53	28	48[a]
Triple negative (vs not triple negative)	8	10[a]	23	10	22[a]
Education					
More than high school (HS) (vs HS or less)	64	57[b]	55	45	44[b]
College degree (vs no college degree)	26	23[b]	26	34	15[b]

Table 4 Demographics, cancer stage, BMI and physical activity: Comparing study population to state-level data for South Carolina.—Cont'd

	EA		AA		
	Current study	State data, %	Current study		
			Non-SI	SI	
	$N=42$, %		$N=66$, %	$N=29$, %	State data, $N=6688$, %
Overweight/obese: (BMI $\geq 25\,kg/m^2$) (vs normal/underweight)	79	61[b]	80	90	84[b]
Physical activity					
Some leisure time activity in past 30 days	81	70[c]	62	79	64[c]
Minutes per week[d,e]					
None	22	32[c]	39	21	37[c]
0–149	37	36[c]	32	39	36[c]
150–299	7	19[c]	21	21	15[c]
300+	34	12[c]	8	19	12[c]
CDC activity guidelines met[d–f]	41	31[c]	29	39	27[c]

[a]Includes all cases of invasive breast cancer of known stage diagnosed throughout South Carolina during 2011–2012, among adult EA ($N=5025$) or AA ($N=1663$) women. Triple Negative analysis excludes cases without complete marker data. Provided by the state central cancer registry.
[b]Education and BMI state data: from South Carolina Behavioral Risk Factor Surveillance Survey (BRFSS) 2012–2013, for EA and AA women ages 37–89 years of age, provided by the state health department.
[c]Physical activity data: from South Carolina Behavioral Risk Factor Surveillance Survey (BRFSS) 2013, for EA and AA women ages 37–89 years of age, provided by the state health department. To be consistent with the current study, BRFSS data were limited to the first reported physical activity only.
[d]Minutes per week: Includes moderate aerobic physical activity, plus vigorous activity where 1 min of vigorous activity counts as 2 min of moderate activity; excludes activity for less than 10 min per session, or of less than 3.0 metabolic equivalents (National Center for Chronic Disease Prevention and Health Promotion. Division of Nutrition Physical Activity and Obesity. CDC, 2013).
[e]Two participants (one EA, one SI) did not provide information as to frequency or duration of physical activity and are excluded from this analysis.
[f]CDC guidelines: at least 150 min per week moderate aerobic physical activity, as defined above.
EA, European American; AA, African American; SI, Sea Islanders; BMI, body mass index; HS, high school; CDC, Centers for Disease Control and Prevention.

3.8 Association between race/ethnicity and triple negative BC subtype

Table 5 presents the percentage of cases with hormone receptor positive BC was significantly lower in non-SI AAs than in the SI or EA racial/ethnic groups ($P < 0.05$). The less-genetically admixed groups (SIs and EAs) had a lower percentage of ER negative BC.

Table 5 ER status, ER/PR/HER2 breast cancer status, and percentage of triple negative breast cancer cases by racial/ethnic group.[a]

	Non-SI AA, $N = 32$ (%)	SI, $N = 24$ (%)	EA, $N = 30$ (%)	*P*-value
ER status				
Missing/indeterminate, *n*	2	0	0	
Number of ER + (% from non-missing)	18 (60)	21 (88)	27 (90)	
	Test for *overall* association among racial groups: $P = 0.01$			
	Non-SI AA vs EA; $P = 0.02$			
	Non-SI AA vs SI; $P = 0.03$			
	EA vs SI; $P = 0.99$			
Breast cancer status	Non-SI AA, %	SI, %	EA, %	
ER status				
Positive	60	88	90	0.01
Negative	40	12	10	
PR status				
Positive	60	96	80	0.01
Negative	40	4	20	
HER2 status				
Positive	28	20	12	0.24
Negative	68	60	65	
Equivocal	4	20	23	
Triple negative, *n* (%)	7 (21.9)	1 (4.2)	2 (6.7)	n/a
Total ($N = 86$), *n* (%)	10 (11.6)			

[a]Missing complete tumor marker data on 51 participants.
ER, estrogen receptor; PR, progesterone receptor; HER2, human epidermal growth factor receptor 2; SI, Sea Islander; AA, African American; EA, European American.

4. Limitations

This exploratory study is one of the first to compare BMI, physical activity, and BC subtypes among the three racially/ethnically diverse population groups. The sample sizes of EAs and SIs were relatively small in comparison to the number of non-SI AAs. However, statistically significant differences in ER positive BC subtype were found even with these relatively small sample sizes. Therefore, the data may have captured real group differences.

The sample was drawn from one state, South Carolina, and the data may not be generalizable to other states. However, the study could be replicated using cancer registry data from other states, to ascertain whether similar results are found.

The study relied on self-reported behavioral data. In future prospective studies, objective measures of physical activity and overweight/obesity could be employed, such as electronic activity tracker data, waist-to-hip ratio, skinfold thicknesses, or bioelectrical impedance.

5. Discussion

This statewide cancer registry-based study of BC survivors has many strengths. First, it shows that the age at initial BC diagnosis was similar for both EAs and AAs. This is also the first known study to examine racial/ethnic differences in BC subtype in a sample that includes the Sea Island/Gullah population (SIs) and is one of the first studies to document that while BMI rates were similar among the three racial/ethnic groups, physical activity rates and BC subtypes were more similar between EAs and SIs as compared to non-SI AAs. The data show that when hormone receptor status was taken into account, similar frequencies were seen in EAs and SI, compared to the frequencies seen in non-SI AAs.

This study is also innovative in evaluating prevalence of localized, regional, and distant BC in the SI/Gullah population. Again, similar frequencies were seen in EAs and SIs than in non-SI AAs.

With regard to BMI, the majority of participants (82%) were overweight/obese regardless of racial/ethnic group. Additionally, BMI was highest in younger non-SI AAs and older participants. Regardless of race/ethnicity, older participants were more likely to report adhering to CDC physical activity guidelines. However, only 28% of participants in the sample met the CDC physical activity guidelines ($\geq$150 min/week). Each major study finding is discussed in greater detail below.

5.1 Relationship between overweight/obesity and BC recurrence

Eighty-two percent of participants in our study had an overweight/obese BMI. This is of great concern as obese women diagnosed with early stage BC following screening mammography have a 2.4-fold higher risk of BC recurrence and death due to BC within 10 years of diagnosis as compared to normal weight women (Kamineni et al., 2013). This association between overweight and obesity and decreased survival is seen both in premenopausal and postmenopausal women.

The results of prior studies have shown that AA women, who have the highest BC mortality rates and levels of overweight/obesity, are ideal targets for physical activity/weight reduction interventions. Therefore, in the future, investigators could focus on developing these interventions with BC survivors, particularly among AA women.

5.2 BC subtype and ethnicity

To explain the racial/ethnic differences in BC subtype observed in this study, it could be hypothesized that European admixture introduced variants affecting immune responses into African genomes. Regulatory variants affecting steady-state gene expression and transcriptional responsiveness to immune challenges may have been preferentially introduced into African genomes through admixture with Europeans primarily through the United States slave trade, which may have conferred a natural selection disadvantage to modern AAs (who, as a group, are only 400 years old, which is extremely young from an evolutionary perspective). This could explain why, for almost every type of cancer, AAs appear to have the worst mortality outcomes of any other racial/ethnic group in the United States (Gaye, Gibbons, Barry, Quarells, & Davis, 2017).

While differential access to health care among AAs is certainly a contributing factor to these worse outcomes, this factor alone cannot fully explain mortality differences, and immune response differences may play a large role in these outcomes. For example, for hormonally driven cancers such as BC, recent research shows that even after controlling for access to care, persistent disparities in survival outcomes exist for AAs vs EAs (Albain, Unger, Crowley, Coltman, & Hershman, 2009; Newman & Kaljee, 2017).

The knowledge gained from these studies could shed light on the associations among ancestral background, the underlying genetic/biomarker distinctions in BC subtypes, and cancer immunologic function. In future studies, targeted dietary and physical activity behavioral interventions could be tested for their effect on reducing the risk of BC recurrence among survivors.

Acknowledgments

This work was supported by United States Department of Defense Southeastern Virtual Institute for Health Equity and Wellness grants W81XWH-11-2-0164 and W81XWH-10-2-0057, United States Department of Defense grant W81XWH-12-1-0043, and by National Institutes of Health grants P20CA157071, R01MD005892, U54CA210962, UL1TR001450, and P30CA138313.

There are no conflicts of interest to report among the authors. The authors thank Ms. Chelsea Lynes of the Division of Surveillance, Office of Public Health Statistics and Information Services, South Carolina Department of Health and Environmental Control for the BRFSS data.

References

Albain, K. S., Unger, J. M., Crowley, J. J., Coltman, C. A., Jr., & Hershman, D. L. (2009). Racial disparities in cancer survival among randomized clinical trials patients of the Southwest Oncology Group. *Journal of the National Cancer Institute*, *101*(14), 984–992. https://doi.org/10.1093/jnci/djp175.

American Cancer Society. (2013). *Cancer facts & figures 2013*. Retrieved from Atlanta, GA https://www.cancer.org/research/cancer-facts-statistics/all-cancer-facts-figures/cancer-facts-figures-2013.html.

American Cancer Society. (2014). *Cancer facts & figures 2014*. Retrieved from Atlanta, GA https://www.cancer.org/research/cancer-facts-statistics/all-cancer-facts-figures/cancer-facts-figures-2014.html.

Campbell, K. L., Van Patten, C. L., Neil, S. E., Kirkham, A. A., Gotay, C. C., Gelmon, K. A., et al. (2012). Feasibility of a lifestyle intervention on body weight and serum biomarkers in breast cancer survivors with overweight and obesity. *Journal of the Academy of Nutrition and Dietetics*, *112*(4), 559–567. https://doi.org/10.1016/j.jada.2011.10.022.

Centers for Disease Control and Prevention (CDC). (2011). *Behavioral Risk Factor Surveillance System survey questionnaire*. Retrieved from https://www.cdc.gov/brfss/questionnaires/index.htm.

Centers for Disease Control and Prevention (CDC). (2015). *Assessing your weight*. Retrieved from https://www.cdc.gov/healthyweight/assessing/index.html.

Centers for Disease Control and Prevention (CDC). (2017). *Physical activity*. Retrieved from https://www.cdc.gov/physicalactivity/resources/recommendations.html.

Centers for Disease Control and Prevention (CDC). (2018). *National Health Interview Survey*. Retrieved from https://www.cdc.gov/nchs/nhis/index.htm.

Demark-Wahnefried, W., Campbell, K. L., & Hayes, S. C. (2012). Weight management and its role in breast cancer rehabilitation. *Cancer*, *118*(Suppl. 8), 2277–2287. https://doi.org/10.1002/cncr.27466.

Fabian, C. J. (2012). Simplifying the energy balance message for breast cancer prevention. *Cancer Prevention Research (Philadelphia, PA)*, *5*(4), 511–514. https://doi.org/10.1158/1940-6207.CAPR-12-0088.

Gaye, A., Gibbons, G. H., Barry, C., Quarells, R., & Davis, S. K. (2017). Influence of socio-economic status on the whole blood transcriptome in African Americans. *PLoS One*, *12*(12), e0187290. https://doi.org/10.1371/journal.pone.0187290.

Kamineni, A., Anderson, M. L., White, E., Taplin, S. H., Porter, P., Ballard-Barbash, R., et al. (2013). Body mass index, tumor characteristics, and prognosis following diagnosis of early-stage breast cancer in a mammographically screened population. *Cancer Causes & Control*, *24*(2), 305–312. https://doi.org/10.1007/s10552-012-0115-7.

McLean, D. C., Jr., Spruill, I., Argyropoulos, G., Page, G. P., Shriver, M. D., & Garvey, W. T. (2005). Mitochondrial DNA (mtDNA) haplotypes reveal maternal population genetic affinities of Sea Island Gullah-speaking African Americans. *American Journal of Physical Anthropology*, *127*(4), 427–438. https://doi.org/10.1002/ajpa.20047.

National Center for Chronic Disease Prevention and Health Promotion. Division of Nutrition Physical Activity and Obesity. CDC. (2013). *A data users guide to the BRFSS physical activity questions*. In *How to assess the 2008 physical activity guidelines for Americans*. Atlanta, GA. Retrieved from https://www.cdc.gov/brfss/pdf/PA%20RotatingCore_BRFSSGuide_508Comp_07252013FINAL.pdf.

Newman, L. A., & Kaljee, L. M. (2017). Health disparities and triple-negative breast cancer in African American women: A review. *JAMA Surgery*, *152*(5), 485–493. https://doi.org/10.1001/jamasurg.2017.0005.

Ogden, C. L., Carroll, M. D., Fryar, C. D., & Flegal, K. M. (2015). *Prevalence of obesity among adults and youth: United States, 2011–2014*. Hyattsville, MD: National Center for Health Statistics. Retrieved from https://www.cdc.gov/nchs/products/databriefs/db219.htm.

South Carolina Department of Health and Environmental Control. (2017). *South Carolina Central Cancer Registry overview*. Retrieved from http://www.scdhec.gov/Health/DiseasesandConditions/Cancer/CancerStatisticsReports/CancerRegistry/.

StataCorp. (2007). *Stata statistical software: Release 10*.

Thompson, C. L., Owusu, C., Nock, N. L., Li, L., & Berger, N. A. (2014). Race, age, and obesity disparities in adult physical activity levels in breast cancer patients and controls. *Frontiers in Public Health*, *2*, 150. https://doi.org/10.3389/fpubh.2014.00150.

Tishkoff, S. A., Reed, F. A., Friedlaender, F. R., Ehret, C., Ranciaro, A., Froment, A., et al. (2009). The genetic structure and history of Africans and African Americans. *Science*, *324*(5930), 1035–1044. https://doi.org/10.1126/science.1172257.

US Cancer Statistics Working Group. (2017). *United States cancer statistics: 1999–2014 incidence and mortality web-based report*. Atlanta, GA: U.S. Department of Health and Human Services, Centers for Disease Control and Prevention and National Cancer Institute. Retrieved from www.cdc.gov/uscs.

Zakharia, F., Basu, A., Absher, D., Assimes, T. L., Go, A. S., Hlatky, M. A., et al. (2009). Characterizing the admixed African ancestry of African Americans. *Genome Biology*, *10*(12), R141. https://doi.org/10.1186/gb-2009-10-12-r141.

CHAPTER FIVE

Race differences in mobility status among prostate cancer survivors: The role of socioeconomic status

Roland J. Thorpe Jr.[a,b,*], **Marino A. Bruce**[c], **Daniel L. Howard**[d], **Thomas A. LaVeist**[a,e]

[a]Program for Research on Men's Health, Hopkins Center for Health Disparities Solutions, Johns Hopkins Bloomberg School of Public Health, Baltimore, MD, United States
[b]Department of Health, Behavior & Society, Johns Hopkins Bloomberg School of Public Health, Baltimore, MD, United States
[c]Program for Research on Faith and Health, Center for Research on Men's Health, Vanderbilt University, Nashville, TN, United States
[d]Public Policy Research Institute and Department of Sociology, Texas A&M University, College Station, TX, United States
[e]Department of Health Policy and Management, Tulane School of Public Health and Tropical Medicine, New Orleans, LA, United States
*Corresponding author: e-mail address: rthorpe@jhu.edu

Contents

Abstract

The objective of this paper was to determine whether there were any race differences in mobility limitation among PCa survivors, and understand the impact of socioeconomic status (SES) on this relationship. Data consisted of 661 PCa survivors (296 Black and 365 White) from the Diagnosis and Decisions in Prostate Cancer Treatment Outcomes (DAD) Study. Mobility limitation was defined as PCa survivors who reported difficulty walking a quarter mile or up 1 flight of stairs. Race was based on the PCa survivors self-identification of either White or Black. SES consisted of education level (i.e., less than high school, high school/GED, some college/associate, bachelors, masters/PhD) and annual household income (i.e., less than $50,000; $50,000–$100,000; greater than $100,000). Adjusting for age, marital status, health insurance, Gleason Score, treatment received, and time to treatment, Black PCa survivors had a higher prevalence of mobility

Advances in Cancer Research, Volume 146
ISSN 0065-230X
https://doi.org/10.1016/bs.acr.2020.01.006

limitation (PR = 1.58, 95% CI: 1.17–2.15) relative to White PCa survivors. When adding education and income to the adjusted model, Black PCa survivors had a similar prevalence of mobility limitation (PR = 1.12, 95% CI: 0.80–1.56) as White PCa survivors. The unequal distribution of SES resources between Black and White PCa survivors accounted for the observed race differences in mobility limitation. This work emphasizes the importance of SES in understanding race differences in mobility among PCa survivors.

1. Introduction

African American men have a higher incidence and death rate of prostate cancer (PCa) than any other racial or ethnic group (Aziz, 2007; Brawley, 2012; Howlader et al., 2016). Although diagnosed at a younger age, African American men tend to present at a later stage of progression of the disease (Aziz, 2007; Brawley, 2012; Howlader et al., 2016). Compared to White men, African American men are more likely to have metastatic disease, and when treated for PCa, lower survival rates (Bellizzi, Mustian, Palesh, & Diefenbach, 2008; Brawley, 2012; Chornokur, Dalton, Borysova, & Kumar, 2011; Eton, Lepore, & Helgeson, 2001; Given, Given, Azzouz, & Stommel, 2001; Lubeck et al., 2001). PCa survival has been linked to health and functional status problems independent of the cancer itself (Litwin et al., 1995; Potosky et al., 1999) and interest in focusing on issues related to quality of life and health-related outcomes has been growing (Deimling, Schaefer, Kahana, Bowman, & Reardon, 2003; Eton et al., 2001; Lubeck et al., 2001). Despite this burgeoning line of research, the studies examining the disparities in mobility status among PCa survivors are virtually nonexistent.

Mobility status represents an indicator of overall underlying health status and age-related decline in physical function among older adults that varies by race (Simonsick et al., 2008; Thorpe, Koster, et al., 2011; Thorpe, Clay, Szanton, Allaire, & Whitfield, 2011). Also mobility status is often considered as a functional limitation—the third stage of the disablement process (Verbrugge & Jette, 1994). This process has been described as a multi-stage pathway beginning with the presence of pathology (i.e., chronic health conditions) and advancing through stages of impairment, functional limitation and, ultimately, disability (Verbrugge & Jette, 1994). Maintaining mobility is an essential component for independence and good quality of life for middle age to older age adults including PCa survivors (Hewitt, Rowland, & Yancik, 2003; Simonsick et al., 2008; Thorpe, Clay, et al., 2011; Thorpe, Koster, et al., 2011). Often defined as reported difficulty walking for one-quarter mile or climbing one flight of stairs (Simonsick et al., 2008; Thorpe, Clay, et al., 2011; Thorpe, Koster, et al., 2011), mobility limitation is an antecedent to mobility disability, which has been associated with adverse health events in older adults (Guralnik, Ferrucci, Simonsick, Salive, & Wallace, 1995; Wolinsky et al., 2007; Wolinsky, Miller, Andresen, Malmstrom, & Miller, 2005). Further, because mobility limitation represents an early stage of age-related decline (Simonsick et al., 2008), there is an opportunity to gain insight into how to prolong mobility limitation for African American and White PCa survivors.

Education and income have been shown to be associated with mobility limitation (Thorpe, Koster, et al., 2011). However, race and socioeconomic status (SES) are often correlated such that it is difficult to tease apart the independent or joint effects of each on health outcomes. Furthermore when examining race differences in mobility limitation, the omission of understanding how education and income influence race differences can lead to spurious conclusions (Thorpe et al., 2014). Prior work examining race differences in disability in PCa survivors fails to consider the impact of SES on race differences in disability (Deimling et al., 2003). Furthermore disability and mobility limitation represent different stages of the disablement process (Verbrugge & Jette, 1994) which are likely to have different relations with race and SES (Thorpe, Koster, et al., 2011; Thorpe, Szanton, Bell, & Whitfield, 2013). As such there is a paucity of research that focuses on race, SES, and mobility among PCa survivors.

Given the increasing number of aged minorities and the increasing number of PCa survivors (Americans, 2016; Howlader et al., 2016; Parry, Kent, Mariotto, Alfano, & Rowland, 2011), a greater focus on understanding race-related disparities in mobility limitation among PCa survivors is needed. The objective of this paper was to determine whether there were any race differences in mobility limitation among PCa survivors, and understand the impact of SES on this relationship. We anticipate that Black PCa survivors would have worse mobility limitation compared to White PCa survivors. However the inclusion of SES variables would eliminate this race difference in mobility limitation.

2. Methods

The Diagnosis and Decisions in Prostate Cancer Treatment Outcomes (DAD) Study is a cross-sectional study designed to examine factors that influence the selection of PCa treatment modality, explore race differences in disease burden, and examine quality of life. We retrospectively recruited 877 PCa survivors (415 Black and 462 White) who were 40–81 years of age that entered the North Carolina Central Cancer Registry (NCCCR) during the years 2007–2008 using a rapid case ascertainment procedure. Eligibility criteria included 35 years of age and older, diagnosed and treated for PCa, and self-identified as White or Black race. Recruitment began in October 2009 and ended in December 2011. NCCCR staff contacted the primary research network hospitals on a monthly basis to request reports identifying patients meeting the eligibility criteria. The NCCCR mailed prospective study participants a pamphlet describing the study, including the one time completion of a survey, and informing them that they may be contacted in the future to participate in a study.

After patient eligibility confirmation was made by our study team, the NCCCR mailed the physician of record of each eligible patient a notification of intent to contact the prospective participant about enrolling in the study. Physicians were given 3 weeks to object to our request to contact their patient. If the physician did not refuse patient contact within 3 weeks, we mailed the eligible patient a packet containing a recruitment letter describing the study, a NCCCR brochure, and a copy of the Institutional Review Board (IRB) approved consent and HIPPA forms. In the recruitment letter, we provided a phone number that prospective participants could call for questions or to decline participation. Interviewers contacted the prospective study participant by telephone, screened them for study eligibility, explained the study, answered questions, and sought their participation. If the patient agreed to participate, the interviewer reviewed the consent form, obtained verbal consent and proceeded with the survey questionnaire. The survey consisted of a series of questions related to PCa, the process of care and their quality of life post treatment that took approximately 15 min to administer. The study was approved by the IRBs of the Johns Hopkins Bloomberg School of Public Health, Department of Defense, and NCCCR. Our sample consisted of 661 PCa survivors (296 Black and 365 White) who had complete information on all variables included in these analyses.

2.1 Measures

Mobility status was based on the participant's response of whether their health or physical problems precluded them from walking a quarter mile (approximately 2–3 blocks) or up 1 flight of stairs (approximately 10 steps).

PCa survivors who reported difficulty walking a quarter mile or up 1 flight of stairs were considered to be have mobility limitation compared to those PCa survivors who did not report any difficulty.

Race, the main independent variable, was based on the PCa survivors's self-identification of either White or Black. A binary variable was created to where Black PCa survivors were coded as 1 and White PCa survivors were coded as 0.

Covariates included age (measured in years), marital status (married vs. not married), education level (less than high school, high school/GED, some college/associate, bachelors, masters/PhD), annual household income (less than $50,000; $50,000–$100,000; greater than $100,000) and health insurance coverage. PCa survivors were asked whether or not they had the following type of health insurance: private health insurance, Medicare, Medicaid, CHAMPUS or CHAMPVA. Those PCa survivors who reported in the affirmative to having any of type of health insurance were considered to have health insurance. Gleason scores were obtained from the pathology reports and were separated into three different categories: Low-grade Cancer (≤ 6), Medium-grade Cancer (7) and High-grade Cancer (8–10). Treatment modality was based on the PCa survivors's report of the following: prostatectomy, radiation therapy, radiation beam, hormone therapy, other treatment, or watchful waiting. Time to treatment was based on the amount of time in months that passed between diagnosis and initial treatment. PCa survivors were asked "In what month and year were you diagnosed with prostate cancer?" They were also asked "In what month and year did you receive your first treatment?" A continuous variable representing time between diagnosis and initial treatment was created by subtracting the time in months that lapsed between diagnosis and initial treatment for each man. A three level variable was created to classify time to treatment as less than 3 months, between 3 and 8 months, and greater than 8 months.

2.2 Analysis

Sample characteristics were described for the total sample using means and standard deviations for continuous variables and frequencies for categorical variables. Student's *t*-test for continuous variables and Chi-Square tests for categorical variables were used to examine mean and proportional differences of the select characteristics of the sample by race (Moore Ds, 1998). Modified Poisson regression models were used to examine the relation between race and mobility limitation adjusted for covariates. Modified Poisson regression model with robust standard errors was specified because the binary nature of the outcome variable was considered to be common (greater than 10%) (McNutt, Wu, Xue, & Hafner, 2003; Thorpe, Parker, Cobb, Dillard, & Bowie, 2017; Zou, 2004). The modeling strategy included four models. The first model tested the bivariate association between race and mobility limitation. The second model tested the association between race and mobility limitation after accounting for age, marital status and health insurance. The third model tested the association after adjusting for variables in model two as well as Gleason score, treatment received, and time to treatment. The fourth model tested the association of education and income on the relationship between race and mobility limitation after adjusting for age, marital status, health insurance, Gleason score, treatment received, and time to treatment. Prevalence ratios (PRs) and corresponding 95% confidence intervals (CI) were used to present findings. White PCa survivors were the reference group in all of our analyses. *P*-values less than 0.05 were considered to be significant. All statistical procedures were performed using Stata statistical software, Version 14.1 (StataCorp LP, College Station, TX).

3. Results

Table 1 shows the distribution of the select characteristics for the DAD study participants for the total sample and by race. Of the 661 PCa survivors included in these analyses 44.7% were Black and the average age was 62.4 ± 7.6 years. Over half of the PCa survivors were married, had health insurance, had a low grade PCa, and had a prostatectomy. Approximately one-quarter of the PCa survivors reported annual household income greater than $100,000 or a Bachelor's degree. Nearly half of the PCa survivors reported their time to treatment was less than 3 months. Twenty percent of the PCa survivors reported mobility limitation. When examining the characteristics by race, Black PCa survivors were younger, less likely to be married, and less likely to have some type of health insurance than White PCa survivors. As it relates to education and income, Black PCa survivors were less likely to have an annual household income greater than $100,000, and less likely to have a Master's/PhD relative to White PCa survivors. Black PCa survivors were less likely to have received a prostatectomy, and less likely for their time to treatment to have been between 3 and 8 months compared to White PCa survivors. Black PCa survivors were

Table 1 Distribution of select characteristics for diagnosis and decisions in prostate cancer treatment outcomes among study participants in the total sample and by race.

Characteristic	Total ($n = 661$)	White ($n = 365$)	Black ($n = 296$)	P-value
Race (%)		55.2	44.7	
Age (mean ± SD)	62.4 ± 7.6	63.1 ± 7.6	61.6 ± 7.5	0.009
Married (%)	78.3	85.7	69.2	<0.001
Health insurance (%)	92.8	96.4	88.5	<0.001
Annual household income (%)				<0.001
Less than $50,0000	34.8	17.5	56.0	
$50,000–$100,000	41.6	46.3	35.8	
Greater than $100,000	23.6	36.1	8.1	
Education (%)				<0.001
Less than High School	10.7	4.6	18.2	
High school/GED	23.1	14.5	33.7	
Some college/associate	23.7	24.6	22.6	
Bachelors	24.2	30.6	16.2	
Masters/PhD	18.1	25.4	9.1	
Gleason score (%)				0.187
Low-grade cancer (≤6)	50.8	53.9	46.9	
Medium-grade cancer (7)	41.9	38.9	45.6	
High-grade cancer (8–10)	7.2	7.1	7.4	
Treatment received (%)				0.004
Prostatectomy	75.0	78.3	70.9	
Radiation beam	12.2	9.3	15.8	
Radiation seed	4.5	4.1	5.0	
Hormone therapy	1.2	0.82	1.6	
Other treatment	4.0	3.0	5.4	
Watchful waiting	2.8	4.3	1.01	
Time to treatment (%)				0.001
Less than 3 months	49.9	50.6	48.9	
3–8 months	43.5	46.0	40.5	
Greater than 8 months	6.5	3.2	10.4	
Mobility limitation (%)	20.7	14.5	28.3	<0.001

more likely to report mobility limitations than White PCa survivors. There were no significant race differences observed between Black and White PCa survivors with respect to their Gleason score.

The association between mobility limitation and race is shown in Table 2. In Model 1, Black PCa survivors had a 95% higher prevalence of mobility limitation (PR = 1.95, 95% CI: 1.43–2.65) compared to White PCa survivors. In model 2 after adjusting for age, marital status, and health insurance, Black PCa survivors had a 84% higher prevalence of mobility limitation (PR = 1.84, 95% CI: 1.34–2.51) than White PCa survivors. Even after accounting for clinical-related variables such as Gleason Score, treatment received, and time to treatment, Black PCa survivors still had a higher prevalence of mobility limitation (PR = 1.58, 95% CI: 1.17–2.15) relative to White PCa survivors. When including education and income in the model 4 which contains all the variables from Model 3, Black PCa survivors had a similar prevalence of mobility limitation (PR = 1.12, 95% CI: 0.80–1.56) as White PCa survivors.

Table 2 Prevalence ratios and 95% confidence intervals for the association between mobility limitation and race in the diagnosis and decisions in prostate cancer treatment outcomes study.

	Model 1	Model 2	Model 3	Model 4
Black race	1.95 (1.43-2.65)	1.84 (1.34–2.51)	1.58 (1.17–2.15)	1.12 (0.80–1.56)
Age		1.04 (1.02–1.06)	1.01 (0.99–1.03)	1.00 (0.98–1.02)
Married		0.59 (0.43–0.80)	0.68 (0.50–0.92)	0.81 (0.61–1.08)
Health insurance		0.66 (0.42–1.06)	0.77 (0.48–1.24)	0.85 (0.55–1.32)
Annual household income				
Less than $50,0000				1.00
$50,000-$100000				0.81 (0.54–1.20)
Greater than $100,000				0.75 (0.37–1.49)

Table 2 Prevalence ratios and 95% confidence intervals for the association between mobility limitation and race in the diagnosis and decisions in prostate cancer treatment outcomes study.—Cont'd

	Model 1	Model 2	Model 3	Model 4
Education				
Less than High School				1.00
High School/GED				0.59 (0.42–0.83)
Some college/associate				0.54 (0.35–0.83)
Bachelors				0.32 (0.18–0.58)
Masters/PhD				0.24 (0.11–0.50)
Gleason score				
Low-grade cancer (≤6)			1.00	1.00
Medium-grade cancer (7)			1.31 (0.95–1.80)	1.26 (0.92–1.72)
High-grade cancer (8-10)			1.46 (0.89–2.41)	1.34 (0.82–2.19)
Treatment received				
Prostatectomy			1.00	1.00
Radiation beam			1.84 (1.27–2.67)	1.68 (1.17–2.41)
Radiation seeds			2.81 (1.73–4.58)	2.48 (1.52–4.05)
Hormone therapy			1.66 (0.58–4.71)	1.60 (0.59–4.28)
Other treatment			2.41 (1.52–3.82)	2.06 (1.33–3.20)
Watchful waiting			0.40 (0.05–2.80)	0.47 (0.07–3.24)
Time to treatment				
Less than 3 months			1.00	1.00
3-8 months			0.79 (0.57–1.08)	0.95 (0.69–1.30)
Greater than 8 months			1.32 (0.87–2.01)	1.39 (0.94–2.07)

4. Discussion

The objective of this paper was to determine if there were any race differences in mobility limitation among PCa survivors. We hypothesized that Black PCa survivors would have worse mobility limitation compared to White PCa survivors, and that SES variables, namely education and income, would eliminate the observed race difference Black PCa survivors reported worse mobility limitation than White PCa survivors until SES was accounted for in the models. This work highlights the importance of SES in understanding race differences in mobility among PCa survivors.

Black PCa survivors exhibited worse mobility limitation than White PCa survivors when accounting for behavioral and demographic variables. However when including SES variables, namely education and income, the observed race difference was no longer significant. This finding is consistent with previous work where Thorpe and colleagues examined the role of SES on race differences in Black and White men and women in the Health, Aging, and Body Composition Study (Thorpe, Koster, et al., 2011). In this well-functioning cohort, education played a significant role in eliminating the difference in mobility status. This work now extends to Black and White prostate cancer survivors. These findings suggest that worse mobility limitation among Black PCa survivors relative to White PCa survivors is driven by differences in education and income.

There are some limitations associated with this study. This study only included Black and White PCa survivors, and we are unaware if these findings would apply to other racial/ethnic groups. Because this is a cross-sectional study design, we are unable to draw causal inferences. The Black and White PCa survivors resided in North Carolina. Therefore the external validity of these findings are limited. The study did not control for chronic conditions which are known to be related to mobility limitation (Thorpe, Clay, et al., 2011) because these were not obtained during data collection. Nevertheless, we controlled for tumor grade and treatment modality. Notwithstanding the limitations, we are unaware of any study that has focused on race differences in mobility limitation among Black and White PCa survivors. Further the study has a sufficient number of Black PCa survivors to provide insights on key variables namely psychosocial factors that are often understudied in this population.

In summary, the unequal distribution of SES resources between Black and White PCa survivors accounted for the observed race differences in mobility limitation. Future work should consider unpacking how education and income are inextricably linked to race differences in mobility limitation among PCa survivors. A prospective study examining the role of education and income on the natural history of mobility limitation among Black and White PCa survivors is sorely needed. Understanding the complex relation among race, SES, and mobility limitation among PCa survivors should remain a key priority in cancer disparities research.

Acknowledgments

Funding for this work was provided in part by the US Department of Defense Grant PC060224, Contract W81XWH-07-01. Investigator(s) were supported, in part, by the US Army Medical Research and Materiel Command, Fort Detrick, MD. Roland J. Thorpe, Jr. is supported by NIA K02AG059140 and NIMHD U54MD000214. Marino A. Bruce is supported by NIA K02AG059140-02S1.

References

Americans, O. (2016). *Key indicators of well-being internet URL*. https://agingstats. gov/docs/ LatestReport. Older-Americans-2016-Key-Indicators-of-WellBeing. pdf [WebCite Cache ID 70rmCzXEh].

Aziz, N. M. (2007). Cancer survivorship research: State of knowledge, challenges and opportunities. *Acta Oncologica*, *46*(4), 417–432.

Bellizzi, K. M., Mustian, K. M., Palesh, O. G., & Diefenbach, M. (2008). Cancer survivorship and aging: Moving the science forward. *Cancer*, *113*(12 Suppl), 3530–3539. https://doi.org/10.1002/cncr.23942.

Brawley, O. W. (2012). Prostate cancer epidemiology in the United States. *World Journal of Urology*, *30*(2), 195–200.

Chornokur, G., Dalton, K., Borysova, M. E., & Kumar, N. B. (2011). Disparities at presentation, diagnosis, treatment, and survival in African American men, affected by prostate cancer. *The Prostate*, *71*(9), 985–997.

Deimling, G. T., Schaefer, M. L., Kahana, B., Bowman, K. F., & Reardon, J. (2003). Racial differences in the health of older-adult long-term cancer survivors. *Journal of Psychosocial Oncology*, *20*(4), 71–94.

Eton, D. T., Lepore, S. J., & Helgeson, V. S. (2001). Early quality of life in patients with localized prostate carcinoma: An examination of treatment-related, demographic, and psychosocial factors. *Cancer*, *92*(6), 1451–1459.

Given, B., Given, C., Azzouz, F., & Stommel, M. (2001). Physical functioning of elderly cancer patients prior to diagnosis and following initial treatment. *Nursing Research*, *50*(4), 222–232.

Guralnik, J. M., Ferrucci, L., Simonsick, E. M., Salive, M., & Wallace, R. B. (1995). Lower extremity function in persons over the age of 70 years as a predictor of subsequent disability. *New England Journal of Medicine*, *332*, 556–561.

Hewitt, M., Rowland, J. H., & Yancik, R. (2003). Cancer survivors in the United States: Age, health, and disability. *The Journals of Gerontology Series A: Biological Sciences and Medical Sciences*, *58*(1), M82–M91.

Howlader, N., Noone, A., Krapcho, M., Miller, D., Bishop, K., Altekruse, S., et al. (2016). *SEER cancer statistics review, 1975–2013*, (p. 19). Bethesda, MD: National Cancer Institute.

Litwin, M. S., Hays, R. D., Fink, A., Ganz, P. A., Leake, B., Leach, G. E., et al. (1995). Quality-of-life outcomes in men treated for localized prostate cancer. *JAMA*, *273*(2), 129–135.

Lubeck, D. P., Kim, H., Grossfeld, G., Ray, P., Penson, D. F., Flanders, S. C., et al. (2001). Health related quality of life differences between black and white men with prostate cancer: Data from the cancer of the prostate strategic urologic research endeavor. *The Journal of Urology*, *166*(6), 2281–2285.

McNutt, L.-A., Wu, C., Xue, X., & Hafner, J. P. (2003). Estimating the relative risk in cohort studies and clinical trials of common outcomes. *American Journal of Epidemiology*, *157*(10), 940–943.

Moore Ds, M. G. P. (1998). *Introduction to the practice of statistics*. New York, NY: W.H. Freeman and Company.

Parry, C., Kent, E. E., Mariotto, A. B., Alfano, C. M., & Rowland, J. H. (2011). Cancer survivors: A booming population. *Cancer Epidemiology and Prevention Biomarkers*, *20*(10), 1996–2005.

Potosky, A. L., Harlan, L. C., Stanford, J. L., Gilliland, F. D., Hamilton, A. S., Albertsen, P. C., et al. (1999). Prostate cancer practice patterns and quality of life: The Prostate Cancer Outcomes Study. *Journal of the National Cancer Institute*, *91*(20), 1719–1724.

Simonsick, E. M., Newman, A. B., Visser, M., Goodpaster, B., Kritchevsky, S. B., Rubin, S., et al. (2008). Mobility limitation in self-described well-functioning older adults: Importance of endurance walk testing. *The Journals of Gerontology. Series A, Biological Sciences and Medical Sciences*, *63*(8), 841–847. 63/8/841 [pii].

Thorpe, R. J., Jr., Clay, O. J., Szanton, S. L., Allaire, J. C., & Whitfield, K. E. (2011). Correlates of mobility limitation in African Americans. *The Journals of Gerontology. Series A, Biological Sciences and Medical Sciences*, *66*(11), 1258–1263.

Thorpe, R. J., Jr., Koster, A., Kritchevsky, S. B., Newman, A. B., Harris, T., Ayonayon, H. N., et al. (2011). Race, socioeconomic resources, and late-life mobility and decline: Findings from the Health, Aging, and Body Composition study. *The Journals of Gerontology. Series A, Biological Sciences and Medical Sciences*, *66*(10), 1114–1123.

Thorpe, R. J., Jr., McCleary, R., Smolen, J. R., Whitfield, K. E., Simonsick, E. M., & LaVeist, T. (2014). Racial disparities in disability among older adults: Finding from the exploring health disparities in integrated communities study. *Journal of Aging and Health*, *26*(8), 1261–1279.

Thorpe, R. J., Jr., Parker, L. J., Cobb, R. J., Dillard, F., & Bowie, J. (2017). Association between discrimination and obesity in African-American men. *Biodemography and Social Biology*, *63*(3), 253–261.

Thorpe, R. J., Jr., Szanton, S. L., Bell, C. N., & Whitfield, K. E. (2013). Education, income and disability in African Americans. *Ethnicity & Disease*, *23*(1), 12–17.

Verbrugge, L. M., & Jette, A. M. (1994). The disablement process. *Social Science & Medicine*, *38*(1), 1–14.

Wolinsky, F. D., Miller, D. K., Andresen, E. M., Malmstrom, T. K., & Miller, J. P. (2005). Further evidence for the importance of subclinical functional limitation and subclinical disability assessment in gerontology and geriatrics. *The Journals of Gerontology. Series B, Psychological Sciences and Social Sciences*, *60*(3), S146–S151.

Wolinsky, F. D., Miller, D. K., Andresen, E. M., Malmstrom, T. K., Miller, J. P., & Miller, T. R. (2007). Effect of subclinical status in functional limitation and disability on adverse health outcomes 3 years later. *The Journals of Gerontology Series A: Biological Sciences and Medical Sciences*, *62*(1), 101–106.

Zou, G. (2004). A modified poisson regression approach to prospective studies with binary data. *American Journal of Epidemiology*, *159*(7), 702–706.

CHAPTER SIX

Assessing an intervention to increase knowledge related to cervical cancer and the HPV vaccine

Marvella E. Ford[a,b,c,*], Kimberly Cannady[d], Georges J. Nahhas[e,f], Kendrea D. Knight[a], Courtney Chavis[a,e], Brittney Crawford[e], Angela M. Malek[a], Erica Martino[a,e], Starr Frazier[g], Antiqua Gathers[g], Claudia Lawton[h], Kathleen B. Cartmell[i], John S. Luque[j]

[a]Professor, Department of Public Health Sciences, Medical University of South Carolina, Charleston, SC, United States
[b]Associate Director, Population Sciences and Cancer Disparities, Hollings Cancer Center, Medical University of South Carolina, Charleston, SC United States
[c]SmartState Endowed Chair in Cancer Disparities Research, South Carolina State University, Orangeburg, SC, United States
[d]Academic Affairs Faculty, Medical University of South Carolina, Charleston, SC, United States
[e]Hollings Cancer Center, Medical University of South Carolina, Charleston, SC, United States
[f]Department of Psychiatry and Behavioral Science, Medical University of South Carolina, Charleston, SC, United States
[g]Department of Biological and Physical Sciences, South Carolina State University, Orangeburg, SC, United States
[h]Institute of Psychiatry, Medical University of South Carolina, Charleston, SC, United States
[i]Department of Public Health Sciences, Clemson University, Clemson, SC, United States
[j]Institute of Public Health, College of Pharmacy and Pharmaceutical Sciences, Florida A&M University, Tallahassee, FL, United States
*Corresponding author: e-mail address: fordmar@musc.edu

Contents

Advances in Cancer Research, Volume 146
ISSN 0065-230X
https://doi.org/10.1016/bs.acr.2020.01.007

Abstract

Human papillomavirus (HPV) infection is the primary risk factor for cervical cancer. While the HPV vaccine significantly reduces the risk of HPV infection and subsequent cervical cancer diagnosis, underuse is linked to lack of knowledge of its effectiveness in preventing cervical cancer. The purpose of this study was to evaluate a cancer educational intervention (titled "MOVENUP") to improve knowledge of cervical cancer, HPV, and the HPV vaccine among predominantly African American communities in South Carolina. The MOVENUP cancer educational intervention was conducted among participants residing in nine South Carolina counties who were recruited by community partners. The 4.5-h MOVENUP cancer educational intervention included a 30-min module on cervical cancer, HPV, and HPV vaccination. A six-item investigator-developed instrument was used to evaluate pre- and post-intervention changes in knowledge related to these content areas. Ninety-three percent of the 276 participants were African American. Most participants reporting age and gender were 50+ years (73%) and female (91%). Nearly half of participants (46%) reported an annual household income <$40,000 and 49% had not graduated from college. Statistically significant changes were observed at post-test for four of six items on the knowledge scale ($P < 0.05$), as compared to pre-test scores. For the two items on the scale in which statistically significant changes were not observed, this was due primarily due to a baseline ceiling effect.

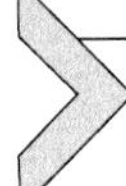

1. Introduction

1.1 Cervical cancer incidence and mortality rates in the United States (US) and in South Carolina

The American Cancer Society notes that 13,170 new cases of invasive cervical cancer were diagnosed in 2019 in the United States, and approximately 4250 women died from cervical cancer (American Cancer Society, 2019). Nationally, the cervical cancer age-adjusted incidence rate from 2008 to 2012 for NHB women was 10.0 per 100,000 reflecting an absolute difference of 2.9 and a rate ratio of 1.41 compared to Non-Hispanic white (NHW) women (American Cancer Society, 2016). African American and Latina women had the highest cervical cancer age-adjusted incidence rates

(8.7 and 9.3 new cases per 100,000 persons, respectively) per year in 2012–2016 compared to white women (7.2 new cases per 100,000 persons) (National Cancer Institute, 2019). During this period, African American and Latina women also had a higher cervical cancer age-adjusted mortality rate (3.5 and 2.6 deaths per 100,000 persons, respectively) per year compared to white women (2.2 deaths per 100,000 persons) (National Cancer Institute, 2019).

According to the most recent state-level data for South Carolina from 2009 to 2017, the age-adjusted cervical cancer mortality rate for NHW women was 2.2 per 100,000 and for African American women was 3.9 per 100,000, for a total of 645 deaths during this time period (South Carolina Department of Health and Environmental Control. Public Health Statistics and Information Services: Division of Biostatistics and Health GIS, 2019b). From 2009 to 2016, the age-adjusted cervical cancer incidence rate for NHW women was lower (7.7 per 100,000) compared to African American women (9.3 per 100,000) (South Carolina Department of Health and Environmental Control. Public Health Statistics and Information Services: Division of Biostatistics and Health GIS, 2019a).

Evidence suggests there are racial and geographic factors which may contribute to the cancer disease burden with higher risk of mortality for underserved, rural women in the southern region of the United States (Moore et al., 2018). African American women and men in South Carolina also experience cancer health disparities for many preventable cancer types (South Carolina Department of Health and Environmental Control. Public Health Statistics and Information Services: Division of Biostatistics and Health GIS, 2019b).

Cervical cancer health disparities have been partially explained by the lack of early detection due to non-adherence to regular screening, and irregular or delayed follow-up after a positive screening test (Musselwhite et al., 2016; Smith et al., 2013). These differences in access to healthcare providers are usually experienced disproportionately by disenfranchised low-income and uninsured individuals. Further, cervical cancer is often described as a disease of poverty (Tsu & Ginsburg, 2017), given that it is almost entirely preventable if patients receive regular cervical cancer screening.

1.2 The importance of the HPV vaccine in preventing cervical Cancer

Human papillomavirus (HPV) is a viral infection, with over 100 different subtypes. Each year, about 14 million people in the United States, including teenagers, are newly infected with HPV (Centers for Disease Control and Prevention, 2017). HPV is most frequently transmitted through skin to skin contact via vaginal, anal, or oral sex. In total, approximately 79 million individuals in the United States are currently living with an HPV infection, most of whom are asymptomatic and do not experience negative health effects (Centers for Disease Control and Prevention, 2017). Therefore, HPV infection is fairly common, and most infections are resolved by the immune system without a person even knowing they were infected.

The progression of persistent HPV infection is the leading known contributor to cervical cancer. However, with proper screening and vaccination, 93% of cervical cancers could be prevented (Centers for Disease Control and Prevention, 2014). The HPV vaccine is recommended for males and females at ages 11–12, along with catch-up vaccination for those up to age 26 who missed vaccination at ages 11–12. The HPV vaccine is also approved by the United States. Food and Drug Administration (FDA) for administration as early as age 9 and up through the age of 45 (Centers for Disease Control and Prevention, 2019). Vaccination at ages 11–12 is optimal because younger adolescents have a strong immune response and the HPV vaccine is most effective prior to HPV exposure (Noronha, Markowitz, & Dunne, 2014).

HPV vaccines can provide protection for most cancer-causing HPV strains and prevent cervical cancer. The original 4-valent HPV vaccine was made from recombinant viral DNA of the L1 capsid protein encompassing strains 6, 11, 16, and 18 of the HPV. The 9-valent HPV vaccine, which is the vaccine used today, is similar but adds coverage to strains 31, 33, 45, 52, and 58 (Lopalco, 2017). HPV strains 16 and 18 are responsible for >80% of cervical cancer diagnoses in the United States; therefore, the HPV vaccine is highly effective in preventing cervical cancer (Lazcano-Ponce et al., 2019). A long-term, 10-year follow-up study of the 4-valent vaccine reported that no cases of HPV types 6, 11, 16, or 18-related diseases were identified (Ferris et al., 2017).

1.3 Disparities in HPV vaccination rates

According to the National Immunization Survey for Teens (NIS-Teen) data, the percentage of 13- to 17-year-old girls who initiated and completed HPV vaccination in 2018 was 69.9% and 53.7%, respectively (Walker et al., 2019). The percentage of 13- to 17-year-old-boys who initiated and completed HPV vaccination was lower at 66.3% and 48.7%, respectively (Walker et al., 2017). In terms of HPV vaccine series completion in the United States by race and ethnicity, the NIS-Teen data showed that

HPV vaccine completion rates range from 47.8% for NHW adolescents to 53.3% for African American adolescents to 56.6% for Hispanic adolescents (Walker et al., 2017).

State level rates for HPV vaccine completion range from a low of 32.6% in Oklahoma to a high of 78.1% in Rhode Island. South Carolina has observed a 4.3% 5-year annual average increase in HPV vaccination coverage, and the HPV vaccine completion rate was 41.2% in 2018. This is below the regional average of 46.7% and the national average of 51.1% (Walker et al., 2019). South Carolina, and other states in the South, remain far behind HPV vaccination rates compared to states in the Northeast and the Pacific Coast.

Past studies have shown that there is a need to improve parent, patient, and provider education on HPV vaccination; reduce vaccine costs; and improve the quality of age-appropriate (11–12 years old) provider recommendations (Gilkey et al., 2016; Lake, Kasting, Christy, & Vadaparampil, 2019). Particularly important for African American patients may be the level of trust in healthcare provider advice. One study reported African American parents who had higher trust in healthcare providers were twice as likely to follow through with HPV vaccine for their children (Fu, Zimet, Latkin, & Joseph, 2017).

1.4 Impact of the lack of knowledge of cervical cancer and/or HPV infection on HPV vaccination uptake

Given the suboptimal rates of HPV vaccination completion and low levels of knowledge of cervical cancer in underserved communities in South Carolina, the investigators of the present study implemented a cancer educational intervention aimed at assessing improvement in knowledge related to cervical cancer, HPV, and the HPV vaccine among intervention participants living in predominantly African American communities. Below we describe the design, methods, and results of the cancer educational intervention to improve knowledge of cervical cancer, HPV, and the HPV vaccine in nine predominantly African American communities in South Carolina.

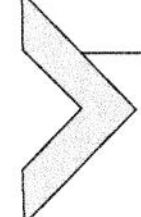

2. Methods

2.1 Study design

A pre-test/post-test design was used to evaluate knowledge changes as a result of the MOVENUP cancer educational intervention. Results related to cancer knowledge outcomes are reported elsewhere and demonstrated that general cancer knowledge and prostate cancer knowledge scores increased significantly after the cancer educational intervention (Ford et al., 2011). Whereas the focus of our prior work reported changes in cancer knowledge, the focus of the current study is to compare pre- and post-intervention changes in knowledge related to cervical cancer, HPV, and HPV vaccination.

2.2 IRB approval

The MOVENUP cancer educational intervention study protocol was approved by the Institutional Review Board (IRB) at the Medical University of South Carolina. Inclusion criteria included:

- Residence in the communities near the location of the intervention site.
- Self-identified gender.
- Self-identified race and ethnicity.
- Ages 21 years or older.

The pre- and post-test intervention surveys were completed by each participant and linked by an identifier that was not connected to his/her name, date of birth, or any other personal identifiers. Therefore, the investigators had no way of connecting survey responses to individual participants in the sessions, so changes in knowledge were evaluated at the group level rather than by individual participants.

2.3 Participant identification and recruitment

The study included participants from communities with large racial disparities in cancer mortality rates (Table 1). Community leaders (Champions) in each county were charged with recruiting participants to each MOVENUP session. While most of the recruited participants were African American, community members from all racial/ethnic groups were eligible to participate.

2.4 MOVENUP cancer educational intervention

The 4.5-h cancer educational intervention (including the cervical cancer and HPV educational module) is called MOVENUP and was delivered in its entirety to participants at each of the 15 study sites between February 2015 and June 2019 (one session per site). The structure and content of the 4.5-h MOVENUP cancer educational intervention were identical across all study sites and with each group of participants. The MOVENUP cancer educational intervention is unique in South Carolina for its focus on African American, rural, and medically underserved communities experiencing cancer health disparities.

Table 1 Age-adjusted cancer mortality rates per 100,000 population for the South Carolina (SC) countries where the MOVENUP cancer educational intervention was conducted, 2017.

County	Non-Hispanic white	African American/black
Beaufort	121.7	178.2
Charleston	147.1	189.1
Clarendon	128.8	191.8
Colleton	180.8	199.1
Greenville	145.7	159.8
Orangeburg	168.1	181.5
Pickens	161.6	248.3
Sumter	160.3	207.4
York	158.3	155.9
All Counties in SC[a]	156.8	186.1
United States[b]	165.4	190.6

[a]Age-adjusted statewide and county rates for South Carolina use the 2000 US Standard population (South Carolina Department of Health and Environmental Control. Public Health Statistics and Information Services: Division of Biostatistics and Health GIS, 2019b).
[b]United States rates are from 2011 to 2016 (Siegel, Miller, & Jemal, 2019).

The 4.5-h evidence-based MOVENUP cancer educational intervention includes a 3-h module focused on general cancer prevention and control information, a 30-min module describing prostate cancer, a 30-min module discussing the importance of cancer clinical trial participation, and a 30-min module focused on cervical cancer, HPV, and the HPV vaccine.

The general cancer knowledge module of the MOVENUP cancer educational intervention was developed by the South Carolina Cancer Alliance, an 800-member, statewide, nonprofit organization with membership from the lay community, cancer survivorship support groups, healthcare organizations, public health associations, and academia. This component highlighted cancer prevention and control information (e.g., lifestyle interventions, cancer screening, early detection, diagnosis, and treatment options). The other modules of the MOVENUP cancer educational intervention were developed by the investigators. All four modules were developed for community audiences with no expert knowledge about cancer.

The MOVENUP cancer educational intervention was designed to be highly interactive and "hands on" rather than merely didactic.

Participants engaged in role play as they practiced sharing the information they learned with others. They also participated in small group activities to review the information that was presented.

The MOVENUP cancer educational intervention employed a train-the-trainer approach in which each participant received a binder that included copies of the materials that were presented during the 4.5-h training session. The binder materials included the most up-to-date, evidence-based information related to cancer screening, early detection, and treatment as well as cervical cancer and HPV vaccine information. The binders also included overhead transparencies, flash drives, and CDs, which allowed the trained facilitators to make use of the different methods to disseminate the information. Additionally, the binders included talking points for each slide. The trained facilitators were encouraged to simply read the materials when they made presentations in their own communities and not add any extra information, to maintain the fidelity of the information and materials.

At the end of each MOVENUP cancer educational intervention session, each trained facilitator signed an agreement to conduct at least two MOVENUP cancer educational intervention training sessions in the next year. The MOVENUP investigators informed the trained facilitators that the future sessions could take place in venues ranging from large group meetings with their religious, civic, and social organizations to small group meetings held in someone's home.

2.5 Rationale for the selected MOVENUP cancer educational intervention counties and demographic characteristics of the selected counties

The MOVENUP cancer educational intervention was conducted at 15 sites in nine different counties representing several different geographic regions of South Carolina (Table 1). These nine counties were: Beaufort, Clarendon, Colleton, Charleston, Greenville, Orangeburg, Pickens, Sumter, and York counties. The study included a sample of participants in counties with high racial disparities in cancer mortality rates as shown in Table 1. Residents of neighboring counties were also invited to participate. The sociodemographic characteristics of the counties in which the MOVENUP cancer educational intervention was conducted are displayed in Table 2.

Table 2 Sociodemographic characteristics of the South Carolina counties receiving the MOVENUP cancer educational intervention.

County name	Population estimates[a]	White	Black/AA	Amer. Indian	Asian	Native Haw. or Pacific Isl.[b]	Two or more races	Hisp./ Latino	Median HH income[c]	Per capita income	Population below the Pov. level
	N	%	%	%	%	%	%	%	USD	USD	%
United States (Reference)	327,167,434	76.5	13.4	1.3	5.9	0.2	2.7	18.3	57,652	31,177	12.3
South Carolina (Reference)	5,084,127	68.5	27.1	0.5	1.8	0.1	1.9	5.8	48,781	26,645	15.4
Beaufort	188,715	77.9	18.2	0.4	1.4	0.1	1.9	11.2	60,603	34,966	10.7
Charleston	405,905	69.2	26.8	0.4	1.8	0.1	1.7	5.1	57,822	35,587	13.3
Clarendon	33,700	50.2	47.5	0.4	0.8	–	1.0	3.2	35,838	20,616	23.2
Colleton	37,660	59.6	37.2	0.9	0.5	0.1	1.7	3.4	34,996	21,059	22.4
Greenville	514,213	76.4	18.5	0.5	2.6	0.1	2.0	9.3	53,739	29,123	12.4
Orangeburg	86,934	34.7	62.2	0.7	1.0	–	1.4	2.4	34,943	19,489	24.4
Pickens	124,937	88.9	7.0	0.3	2.1	–	1.8	3.8	45,332	23,501	15.3
Sumter	106,512	47.9	47.9	0.6	1.5	0.1	2.0	4.1	41,946	21,733	19.1
York	274,118	75.0	19.4	0.9	2.5	0.1	2.2	5.7	59,394	30,387	11.2

[a]Note: Population estimates, July 2018.
[b]Missing data for some of these values due to small sample size.
[c]Median Household Income/Per Capita income (in 2017 USD), 2013–2017.
Abbreviations: AA, African American; Amer., American; Haw., Hawaiian; Isl., Islander; Hisp., Hispanic; HH, household; Pov., Poverty; USD, United States Dollars.
Source: https://www.census.gov/quickfacts/fact/table/US,colletoncountysouthcarolina/PST045218, accessed on 8-6-2019

2.6 Conceptual framework of the MOVENUP cancer educational intervention

The design of the MOVENUP cancer educational intervention, which includes a cervical cancer and HPV vaccine module, was based on a conceptual framework developed to understand barriers impacting the participation of diverse populations in clinical trials (Swanson & Ward, 1995). These same barriers could negatively impact the likelihood of African American community members participating in educational forums focused on cervical cancer education. The conceptual framework for our study is shown in Fig. 1.

In the Swanson and Ward (1995) framework, sociocultural barriers are defined as fear and mistrust of research, the investigators conducting the research, or even the institution where the research is conducted (Swanson & Ward, 1995). Sociocultural barriers also include racial and ethnic discrimination, cultural beliefs regarding illness and disease, mistrust of the health care system, and differences in health beliefs and practices (Swanson & Ward, 1995). The sociocultural barriers identified in the Swanson and Ward (1995) framework could deter African American community members from receiving the HPV vaccine. For example, African American parents may mistrust the health care system's reassurances that the HPV vaccine is safe and effective (Fu et al., 2017).

In the present study, sociocultural barriers were addressed using the following methods. First, the MOVENUP cancer educational intervention was conducted in trusted community venues including churches, libraries, thrift stores, a community action agency, and a cancer center (see Table 3). Second, the investigators also worked with trusted community partners ("Champions," e.g., civic, religious, and social organizational leaders) who endorsed the study and recruited participants to attend each MOVENUP cancer educational intervention session. Using community-engaged recruitment methods helped to foster a sense of trust among the study participants.

Fig. 1 Conceptual framework.

Table 3 Location, date, and number of participants in the MOVENUP cancer educational intervention by study site ($n = 276$).[a]

SC city ($n = 10$)	Urban/ rural	SC county ($n = 9$)	Study site ($n = 15$)	Date of the intervention	Preregistered participants ($n = 358$)	Intervention participants ($n = 274$)[b]
Rock Hill	Urban	York	Mt. Hebron Baptist Church	February 21, 2015	25	22
St. Helena Island	Rural	Beaufort	St. Helena Island Branch Library	April 16, 2015	15	9
Orangeburg	Rural	Orangeburg	OCAB Community Action Agency	May 23, 2015	25	20
Charleston	Urban	Charleston	MUSC Hollings Cancer Center	June 6, 2015	26	17
Greenville	Urban	Greenville	Buncombe Street United Methodist	August 27, 2016	27	19
Greenville	Urban	Greenville	Goodwill Industries Thrift Store	October 22, 2016	22	11
Greenville	Urban	Greenville	Nicholtown Baptist Church	June 23, 2018	32	27
Pickens	Rural	Pickens	Griffin Ebenezer Baptist Church	July 14, 2018	25	27
Greenville	Urban	Greenville	Long Branch Baptist Church	July 14, 2018	22	10
Greenville	Urban	Greenville	Goodwill Industries Thrift Store	October 6, 2018	19	13

Continued

Table 3 Location, date, and number of participants in the MOVENUP cancer educational intervention by study site ($n = 276$).[a]—Cont'd

SC city ($n = 10$)	Urban/ rural	SC county ($n = 9$)	Study site ($n = 15$)	Date of the intervention	Preregistered participants ($n = 358$)	Intervention participants ($n = 274$)[b]
Charleston	Urban	Charleston	Morris Brown AME Church	January 19, 2019	32	34
Edisto Island	Rural	Charleston	Allen AME Church	March 2, 2019	21	12
Summerton	Rural	Clarendon	Historic Liberty Hill AME Church	April 27, 2019	25	25
Holly Hill	Rural	Orangeburg	Greater Unity AME Church	May 25, 2019	21	12
Wedgefield	Rural	Sumter	Orangehill AME Church	June 28, 2019	21	16

[a]The response rate was 77% for the intervention participants *versus* preregistered participants.
[b]Two participants left the training before completing the survey.
Abbreviations: SC, South Carolina; OCAB, Orangeburg-Calhoun-Allendale-Bamberg; MUSC, Medical University of South Carolina; AME, African Methodist Episcopal.

Additionally, each MOVENUP cancer educational intervention session took place in a trusted community venue that was identified and arranged by trusted local community leaders (Champions) in each community setting. The Champions were responsible for recruiting the participants for the MOVENUP cancer educational intervention trainings. Moreover, many of the investigative team members were African American, and therefore may have been more relatable to the study participants. Finally, as part of the MOVENUP cancer educational intervention, past clinical trial and ethical violations and abuses documented in the Tuskegee Syphilis Study and the Henrietta Lacks cell line controversy were presented to explain how research protections eventually led to the current high levels of protection of participants in contemporary research studies (e.g., the informed consent process, consent forms, and Data and Safety Monitoring Committees).

2.7 Measurement instrument

The study outcome was defined as a significant pre-test/post-test change in knowledge related to cervical cancer, HPV, and the HPV vaccine. A six-item investigator-developed instrument was used to evaluate these changes in knowledge. The questions were based on the investigators' review of the contemporary literature on the topic of cervical cancer and HPV infection and included a mix of True/False and multiple choice questions (see Table 4). An overall knowledge score was also calculated that ranged from 0 to 6 for the total of the six questions, based on one point per each correctly answered question.

- Two items assessed cervical cancer knowledge,
- Two items evaluated HPV knowledge, and
- Two items assessed HPV vaccination knowledge.

2.8 Statistical analysis

Frequencies and percentages were reported for demographic variables. Pre- and post-intervention changes in cervical cancer, HPV, and HPV vaccine knowledge were assessed by McNemar's test of agreement for paired data. Regression analyses assessed the associations of demographic variables (e.g., education, income) and the overall knowledge score. SAS Version 9.4 was used to perform analyses (SAS Institute Inc., Cary, NC).

Table 4 Study instrument assessing cervical cancer, HPV, and HPV vaccination knowledge.

Quiz evaluating cervical cancer, HPV, and HPV vaccination knowledge

Question 1. A Papanicolaou (Pap) test is a screening test for (circle one answer, a–e): a. Breast Cancer b. Colorectal Cancer c. Prostate Cancer d. Cervical Cancer e. None of the Above	Question 4. HPV is transmitted through (circle one answer, a–c): a. Skin-to-skin sexual contact b. Shaking hands with an HPV infected person c. Coming in contact with the blood of an HPV-infected person
Question 2. What is the major cause of cervical cancer (circle one answer, a–d): a. Hepatitis A b. Human Immunodeficiency Virus c. Human Papilloma Virus d. Influenza	Question 5. Only females can receive the HPV vaccine (circle one answer, True or False): a. True b. False
Question 3. Having an HPV infection means that you will get cervical cancer (circle one answer, True or False): a. True b. False	Question 6. Only sexually active individuals can receive the HPV vaccine (circle one answer, True or False): a. True b. False

3. Results

Study sites included a number of churches, one library, two thrift stores, one community action agency, and a cancer center. The number of participants at each site varied from 9 to 34, with an average of 18 participants per site (Table 3).

Table 5 provides a summary of the demographic characteristics of the 276 study participants (93% African American). Most of the participants (73%) who reported age were 50 years of age or older. Among those who reported gender and income, 91% were female and 46% had an annual household income <$40,000. Forty-nine percent of the participants who reported their educational level had not graduated from college.

Table 6 reports the outcomes of the cervical cancer questions. Overall, mean (SD) pre- (4.18 [1.42]) and post-test (4.68 [1.55]) scores differed significantly ($P < 0.0001$). For questions 2, 4, 5, and 6 there was a statistically

Table 5 Summary of demographic characteristics of participants at pre-test ($n = 276$).

Variable	n (%)
Age (years)[a] ($n = 270$)	
<50	74 (27.4)
51–64	94 (34.8)
65+	102 (37.8)
Gender[a] ($n = 161$)	
Male	15 (9.3)
Female	146 (90.7)
Hispanic ethnicity (Yes/No)[a] ($n = 274$)	
Yes	3 (1.1)
No	271 (98.9)
Race[a] ($n = 274$)	
African American or Black	255 (93.1)
White	17 (6.2)
Other[b]	2 (0.8)
Education[a] ($n = 272$)	
Less than high school	9 (3.34)
High school—some college	125 (45.9)
College graduate or more	138 (50.8)
Marital status[a] ($n = 270$)	
Married or living as married	114 (42.2)
Widowed	39 (14.4)
Divorced/Separated	55 (20.4)
Never married	62 (23.0)
Household income (USD) ($n = 261$)	
0–19,999	57 (21.8)
20,000–39,999	63 (24.1)
40,000–59,999	65 (25.0)
60,000–79,999	29 (11.1)
≥80,000	47 (18.0)

[a]Missing data on this variable.
[b]Other race included one individual of American Indian or Alaska Native race.

Table 6 Pre-test/post-test knowledge response changes by question.

Questions (correct response)	Incorrect to incorrect *n* (%)	Incorrect to correct *n* (%)	Correct to incorrect *n* (%)	Correct to correct *n* (%)	*P* value[a]
Q1: A Papanicolaou (Pap) test is a screening test for … (cervical cancer)	20 (7.87)	31 (12.20)	23 (9.06)	180 (70.87)	0.2763
Q2: What is the major cause of cervical cancer? (human papillomavirus)	33 (12.99)	53 (20.87)	16 (6.30)	152 (59.84)	<0.0001
Q3: Having an HPV infection means that you will get cervical cancer. (False)	28 (11.02)	34 (13.39)	38 (14.96)	154 (60.63)	0.6374
Q4: HPV is transmitted through … (skin-to-skin sexual contact)	61 (24.02)	101 (39.76)	9 (3.54)	83 (32.68)	<0.0001
Q5: Only females can receive the HPV vaccine. (False)	17 (6.69)	45 (17.72)	27 (10.63)	165 (64.96)	0.0339
Q6: Only sexually active individuals can receive the HPV vaccine. (False)	18 (7.09)	21 (8.27)	41 (16.14)	174 (68.5)	0.0111

[a]Statistical significance indicated by $P < 0.05$.

significant difference between the pre- and post-test scores ($P < 0.05$). More participants changed from an incorrect to a correct answer for questions 2 and 4, with the greatest change in question 4 (40%) about how HPV was transmitted (skin-to-skin contact). In examining the item analysis, most participants knew the correct answer to question 1, that a Pap test is a screening test for cervical cancer, so there was no change in knowledge at post-test. Likewise, for question 3, most participants knew that having an HPV infection does not necessarily mean a person will develop cervical cancer. At post-test, the average correct response percentage for the 6 questions was 79% (range 74%–83%). In regression models, the demographic variables assessed (age, education, income, urban/rural) were not found to be predictors for the overall knowledge score.

4. Discussion

The MOVENUP cancer educational intervention implemented community-based cancer education in nine counties of South Carolina to engage predominantly African American communities. During each MOVENUP session led by the investigators, important cancer educational topics relevant to African American cancer mortality rates were highlighted.

The results demonstrated that community-delivered cancer education can increase knowledge of cervical cancer in African American communities, and hopefully lead to greater uptake of the HPV vaccine. Given the suboptimal rates of HPV vaccination in South Carolina and disproportionate cervical cancer mortality rates for African American women, there have been statewide efforts to improve HPV vaccination rates and increase adherence to cervical cancer screening through the South Carolina Cancer Alliance and partner institutions (South Carolina Cancer Alliance, 2019). The state cancer plan has a modest goal to increase the HPV vaccination completion rates for 13- to 17-year olds to 50% by the end of 2021 (South Carolina Cancer Alliance, 2019). The MOVENUP cancer educational intervention complements these efforts with its focus on community-engaged educational approaches in African American communities with limited access to cancer educational information. The MOVENUP investigators, with funding from the National Cancer Institute, are now developing strategies to link their educational efforts with no-/low-cost HPV vaccination sites in each county.

Investigators from academic institutions in South Carolina are committed to conducting community-engagement strategies to reduce the glaring health disparities in the state. For example, the National Institutes of Health (NIH) Report, Principles of Community Engagement, highlights the community engagement work in health-related research in South Carolina (Clinical and Translational Science Awards Consortium Community Engagement Key Function Committee Task Force on the Principles of Community Engagement, 2011; Wilcox et al., 2007). One example was the Health-e-AME (African Methodist Episcopal) faith-based physical activity initiative. The community identified physical activity as an intervention target to decrease health disparities. This initiative partnered the AME church with the Medical University of South Carolina and the University of South Carolina to increase physical activity in their congregations.

Following the initiative, there was high awareness of the intervention among parishioners, and the partnership with the AME church continues to collaborate on public health initiatives.

Another highly recognized academic-community research partnership in South Carolina was Project SuGAR, in which African Americans living in the South Carolina Sea Islands partnered with investigators at the Medical University of South Carolina to research genetic factors involved in diabetes and to engage the community in health education and health screenings for metabolic and cardiovascular diseases (Clinical and Translational Science Awards Consortium Community Engagement Key Function Committee Task Force on the Principles of Community Engagement, 2011; Hunt et al., 2014). The Project SuGAR Citizen Advisory Committee was established in 1996 and continues to function today and support other community health research projects studying systemic lupus erythematosus and cancer disparities (Spruill et al., 2013; Wolf et al., 2018). These few examples highlight some of the long-term, community-engaged work by academic institutions in South Carolina, and the current study contributes to this literature as another successful example from the field. As in the other examples cited, the MOVENUP cancer educational intervention employed community engagement principles to improve program outcomes specifically related to access to up-to-date cancer educational resources.

The MOVENUP cancer educational intervention successfully built trust with several predominantly African American communities in rural and urban areas of South Carolina and delivered cancer educational interventions in trusted locations to engaged community members. Examples of community partners included leaders from the following organizations: the 7th Episcopal District of the AME Church of South Carolina, the South Carolina Cancer Alliance, and the American Cancer Society. By partnering with the community, the MOVENUP cancer educational intervention created learning opportunities in cancer education to improve cancer knowledge in these communities.

Other studies aiming to increase HPV vaccination knowledge in African American communities have used a community forum approach, which has also been used in community-engaged research on various health topics (Bharmal et al., 2016). For example, Teteh and colleagues reported a significant change in perceived knowledge of HPV and cervical cancer for both African Americans and Hispanics following the delivery of community forums (Teteh et al., 2019). In the analysis of the post-test scores, the authors found that having a regular doctor and trust of vaccines were associated with higher knowledge scores (Teteh et al., 2019). Survey studies with African American communities have found varied results with knowledge of cervical cancer, HPV and the HPV vaccine. A survey study in Chicago with African American women reported that 73% of participants scored less than 65% on the knowledge part of the survey. Variables associated with higher scores (>65%) included education, income, and having a child vaccinated with the HPV vaccine (Strohl et al., 2015). This study was more similar to the present study in that it studied middle-aged women and found low knowledge scores among community members. In the present study, the average correct post-test knowledge score was 78%.

Another survey study conducted in Charleston, South Carolina with Latina immigrant women focused on cervical cancer screening adherence reported that slightly more than half of the sample had heard of the HPV vaccine and two-thirds believed HPV was a sexually transmitted disease (Luque et al., 2018). In this study, factors negatively associated with cervical cancer screening included not knowing where to go to be screened, not having a regular provider and psychosocial factors such as low self-efficacy. What these other studies have in common with the present study is a demonstrated need for not only cervical cancer education in minority communities, but also additional support for healthcare seeking and addressing the challenges of finding a source of regular healthcare.

In summary, while the goal of this study was to evaluate the proportion of people who answered correctly between the pre- and post-test, the study data appear to be trending toward significant improvements in knowledge, particularly for survey items 2 and 4. The post-test results indicated that participants increased their knowledge about cervical cancer, HPV, and the HPV vaccine.

4.1 Strengths and limitations

There were both strengths and limitations of this research. While the sample was representative of counties with high rates of cervical cancer deaths, a random sampling selection scheme was not used. This could potentially limit the generalizability of the study findings. While the entire state of South Carolina is considered the catchment area of the Medical University of South Carolina Hollings Cancer Center, the MOVENUP cancer educational and outreach intervention focused on counties with sizable African American communities, and the intervention was conducted at only one time point. It is possible that the gains in knowledge related to cervical cancer, HPV, and the HPV vaccine may not be sustained over time. Booster or refresher sessions may be needed in the future to continue to engage these communities and sustain efforts. However, there were many strengths in the academic-community engagement efforts which empowered the community and increased trust in cancer educational research.

Most of the participants were women, who typically serve as the gatekeepers for their family's health activities. Our team of investigators has found that women share the health information they gain with their families, including male relatives. Nevertheless, it will be important in the future to make a concerted effort to include more men in the MOVENUP cancer educational intervention. Changes in behavior related to the HPV vaccination were not measured. In a future study, the investigators will use federal and state data sources to evaluate changes in the pre- and post-intervention HPV vaccination rates in the nine counties where the intervention was conducted.

5. Conclusions

The purpose of this study was to evaluate the results of a cancer educational intervention to improve cervical cancer knowledge among predominantly African American communities in South Carolina. The study results show that overall, the MOVENUP cancer educational intervention led to significant increases in knowledge related to cervical cancer, HPV, and the HPV vaccine. This is especially important for content with which the participants may not have previously been familiar. However, for a few of the items on the study assessment instrument, no significant changes were seen from pre- to post-test. This was largely due to the fact that most participants responded with correct answers at pre-test, leaving little room to improve. As noted above, to measure the behavioral impact of the MOVENUP cancer educational intervention, the investigators will analyze federal and state data to evaluate changes in HPV vaccination uptake rates in the nine counties where the intervention was conducted.

Acknowledgments

NIH/NCI R25 CA193088—South Carolina Cancer Health Equity Consortium (SC CHEC): Summer Undergraduate Research Training Program.

NIH/NCI P30CA138313-10S2—Medical University of South Carolina Cancer Center Support Grant Administrative Supplement to Strengthen NCI Supported Community Outreach Capacity through Community Health Educators (CHEs) of the National Outreach Network (NON).

NIH/NCI P20CA157071—South Carolina Cancer Disparities Research Center. NIH/NIMHD R01MD005892—Improving Resection Rates among African Americans with NSCLC, NIH/NCI U54CA210962—South Carolina Cancer Disparities Research Center, NIH/NCATS.

UL1TR001450—South Carolina Clinical & Translational Research Institute (SCTR), and P30CA138313—Medical University of South Carolina/Cancer Center Support Grant.

The current study incorporated a modified version of a cancer educational intervention that was developed by the South Carolina Cancer Alliance for use by general audiences. This educational intervention which initially focused on general knowledge of cancer was expanded to also include knowledge of cancer clinical trials, cervical cancer (including HPV and HPV vaccination), and prostate cancer. These three components were added to the educational intervention due to disparities in cancer incidence and mortality.

References

American Cancer Society. (2016). *Cancer facts & figures for African Americans 2016–2018*. Retrieved from Atlanta, GA https://www.cancer.org/content/dam/cancer-org/research/cancer-facts-and-statistics/cancer-facts-and-figures-for-african-americans/cancer-facts-and-figures-for-african-americans-2016-2018.pdf.

American Cancer Society. (2019). *Cancer facts & figures for African Americans 2019–2021*. Retrieved from Atlanta, GA https://www.cancer.org/content/dam/cancer-org/research/cancer-facts-and-statistics/cancer-facts-and-figures-for-african-americans/cancer-facts-and-figures-for-african-americans-2019-2021.pdf.

Bharmal, N., Lucas-Wright, A. A., Vassar, S. D., Jones, F., Jones, L., Wells, R., et al. (2016). A community engagement symposium to prevent and improve stroke outcomes in diverse communities. *Progress in Community Health Partnerships*, *10*(1), 149–158. https://doi.org/10.1353/cpr.2016.0010.

Centers for Disease Control and Prevention. (2014). *CDC vital signs*. Retrieved from https://www.cdc.gov/vitalsigns/pdf/2014-11-vitalsigns.pdf.

Centers for Disease Control and Prevention. (2017). *Genital HPV infection—CDC fact sheet*. Retrieved from https://www.cdc.gov/std/hpv/HPV-FS-print.pdf.

Centers for Disease Control and Prevention. (2019). *Immunization schedules*. Retrieved from https://www.cdc.gov/vaccines/schedules/hcp/imz/child-adolescent.html#note-hpv.

Clinical and Translational Science Awards Consortium Community Engagement Key Function Committee Task Force on the Principles of Community Engagement. (2011). *Principles of community engagement* [11-7782] National Institutes of Health. Retrieved from https://www.atsdr.cdc.gov/communityengagement/pdf/PCE_Report_508_FINAL.pdf.

Ferris, D. G., Samakoses, R., Block, S. L., Lazcano-Ponce, E., Restrepo, J. A., Mehlsen, J., et al. (2017). 4-valent human papillomavirus (4vHPV) vaccine in preadolescents and adolescents after 10 years. *Pediatrics*, *140*(6). https://doi.org/10.1542/peds.2016-3947. pii: e20163947.

Ford, M. E., Wahlquist, A. E., Ridgeway, C., Streets, J., Mitchum, K. A., Harper, R. R., Jr., et al. (2011). Evaluating an intervention to increase cancer knowledge in racially diverse communities in South Carolina. *Patient Education and Counseling*, *83*(2), 256–260. https://doi.org/10.1016/j.pec.2010.05.028.

Fu, L. Y., Zimet, G. D., Latkin, C. A., & Joseph, J. G. (2017). Associations of trust and healthcare provider advice with HPV vaccine acceptance among African American parents. *Vaccine*, *35*(5), 802–807. https://doi.org/10.1016/j.vaccine.2016.12.045.

Gilkey, M. B., Calo, W. A., Moss, J. L., Shah, P. D., Marciniak, M. W., & Brewer, N. T. (2016). Provider communication and HPV vaccination: The impact of recommendation quality. *Vaccine*, *34*(9), 1187–1192. https://doi.org/10.1016/j.vaccine.2016.01.023.

Hunt, K. J., Kistner-Griffin, E., Spruill, I., Teklehaimanot, A. A., Garvey, W. T., Sale, M., et al. (2014). Cardiovascular risk in Gullah African Americans with high familial risk of type 2 diabetes mellitus: Project SuGAR. *Southern Medical Journal, 107*(10), 607–614. https://doi.org/10.14423/SMJ.0000000000000172.

Lake, P. W., Kasting, M. L., Christy, S. M., & Vadaparampil, S. T. (2019). Provider perspectives on multilevel barriers to HPV vaccination. *Human Vaccines & Immunotherapeutics, 15*(7–8), 1784–1793. https://doi.org/10.1080/21645515.2019.1581554.

Lazcano-Ponce, E., Torres-Ibarra, L., Cruz-Valdez, A., Salmeron, J., Barrientos-Gutierrez, T., Prado-Galbarro, J., et al. (2019). Persistence of immunity when using different human papillomavirus vaccination schedules and booster-dose effects 5 years after primary vaccination. *The Journal of Infectious Diseases, 219*(1), 41–49. https://doi.org/10.1093/infdis/jiy465.

Lopalco, P. L. (2017). Spotlight on the 9-valent HPV vaccine. *Drug Design, Development and Therapy, 11*, 35–44. https://doi.org/10.2147/DDDT.S91018.

Luque, J. S., Tarasenko, Y. N., Li, H., Davila, C. B., Knight, R. N., & Alcantar, R. E. (2018). Utilization of cervical cancer screening among Hispanic immigrant women in coastal South Carolina. *Journal of Racial and Ethnic Health Disparities, 5*(3), 588–597. https://doi.org/10.1007/s40615-017-0404-7.

Moore, J. X., Royston, K. J., Langston, M. E., Griffin, R., Hidalgo, B., Wang, H. E., et al. (2018). Mapping hot spots of breast cancer mortality in the United States: Place matters for Blacks and Hispanics. *Cancer Causes & Control, 29*(8), 737–750. https://doi.org/10.1007/s10552-018-1051-y.

Musselwhite, L. W., Oliveira, C. M., Kwaramba, T., de Paula Pantano, N., Smith, J. S., Fregnani, J. H., et al. (2016). Racial/ethnic disparities in cervical cancer screening and outcomes. *Acta Cytologica, 60*(6), 518–526. https://doi.org/10.1159/000452240.

National Cancer Institute. (2019). *SEER cancer stat facts: Cervical cancer.* Retrieved from https://seer.cancer.gov/statfacts/html/cervix.html.

Noronha, A. S., Markowitz, L. E., & Dunne, E. F. (2014). Systematic review of human papillomavirus vaccine coadministration. *Vaccine, 32*(23), 2670–2674. https://doi.org/10.1016/j.vaccine.2013.12.037.

Siegel, R. L., Miller, K. D., & Jemal, A. (2019). Cancer statistics, 2019. *CA: A Cancer Journal for Clinicians, 69*(1), 7–34. https://doi.org/10.3322/caac.21551.

Smith, J. S., Brewer, N. T., Saslow, D., Alexander, K., Chernofsky, M. R., Crosby, R., et al. (2013). Recommendations for a national agenda to substantially reduce cervical cancer. *Cancer Causes & Control, 24*(8), 1583–1593. https://doi.org/10.1007/s10552-013-0235-8.

South Carolina Cancer Alliance. (2019). *Services and initiatives, cervical cancer.* Retrieved from https://www.sccancer.org/initiatives/cervical-cancer/.

South Carolina Department of Health and Environmental Control. Public Health Statistics and Information Services: Division of Biostatistics and Health GIS. (2019a). *SCAN cancer incidence.* Full (Research) File. Retrieved from http://scangis.dhec.sc.gov/scan/cancer2/fullinput.aspx.

South Carolina Department of Health and Environmental Control. Public Health Statistics and Information Services: Division of Biostatistics and Health GIS. (2019b). *SCAN cancer mortality.* Retrieved from http://scangis.dhec.sc.gov/scan/cancer2/mortinput.aspx.

Spruill, I. J., Leite, R. S., Fernandes, J. K., Kamen, D. L., Ford, M. E., Jenkins, C., et al. (2013). Successes, challenges and lessons learned: Community-engaged research with South Carolina's Gullah population. *Gateways, 6.* https://doi.org/10.5130/ijcre.v6i1.2805.

Strohl, A. E., Mendoza, G., Ghant, M. S., Cameron, K. A., Simon, M. A., Schink, J. C., et al. (2015). Barriers to prevention: Knowledge of HPV, cervical cancer, and HPV vaccinations among African American women. *American Journal of Obstetrics and Gynecology, 212*(1), 65. e61-65. https://doi.org/10.1016/j.ajog.2014.06.059.

Swanson, G. M., & Ward, A. J. (1995). Recruiting minorities into clinical trials: Toward a participant-friendly system. *Journal of the National Cancer Institute, 87*(23), 1747–1759. https://doi.org/10.1093/jnci/87.23.1747.

Teteh, D. K., Dawkins-Moultin, L., Robinson, C., LaGroon, V., Hooker, S., Alexander, K., et al. (2019). Use of community forums to increase knowledge of HPV and cervical cancer in African American communities. *Journal of Community Health, 44*(3), 492–499. https://doi.org/10.1007/s10900-019-00665-2.

Tsu, V. D., & Ginsburg, O. (2017). The investment case for cervical cancer elimination. *International Journal of Gynaecology and Obstetrics, 138*(Suppl. 1), 69–73. https://doi.org/10.1002/ijgo.12193.

Walker, T. Y., Elam-Evans, L. D., Singleton, J. A., Yankey, D., Markowitz, L. E., Fredua, B., et al. (2017). National, regional, state, and selected local area vaccination coverage among adolescents aged 13-17 years—United States, 2016. *MMWR. Morbidity and Mortality Weekly Report, 66*(33), 874–882. https://doi.org/10.15585/mmwr.mm6633a2.

Walker, T. Y., Elam-Evans, L. D., Yankey, D., Markowitz, L. E., Williams, C. L., Fredua, B., et al. (2019). National, regional, state, and selected local area vaccination coverage among adolescents aged 13–17 years—United States, 2018. *MMWR. Morbidity and Mortality Weekly Report, 68*(33), 718–723. https://doi.org/10.15585/mmwr.mm6833a2.

Wilcox, S., Laken, M., Anderson, T., Bopp, M., Bryant, D., Carter, R., et al. (2007). The health-e-AME faith-based physical activity initiative: Description and baseline findings. *Health Promotion Practice, 8*(1), 69–78. https://doi.org/10.1177/1524839905278902.

Wolf, B. J., Ramos, P. S., Hyer, J. M., Ramakrishnan, V., Gilkeson, G. S., Hardiman, G., et al. (2018). An analytic approach using candidate gene selection and logic Forest to identify gene by environment interactions (G × E) for systemic lupus erythematosus in African Americans. *Genes (Basel), 9*(10). https://doi.org/10.3390/genes9100496. pii: E496.

CHAPTER SEVEN

Patient barriers to cancer clinical trial participation and navigator activities to assist

Kathleen B. Cartmell[a,b,*], Heather S. Bonilha[c], Kit N. Simpson[c], Marvella E. Ford[d,e,f], Debbie C. Bryant[a,b], Anthony J. Alberg[g]

[a]Department of Public Health Sciences, Clemson University, Clemson, SC, United States
[b]Medical University of South Carolina, Hollings Cancer Center, Charleston, SC, United States
[c]Medical University of South Carolina, College of Health Professions, Charleston, SC, United States
[d]Professor, Department of Public Health Sciences, Medical University of South Carolina, Charleston, SC, United States
[e]Associate Director, Population Sciences and Cancer Disparities, Hollings Cancer Center, Medical University of South Carolina, Charleston, SC United States
[f]SmartState Endowed Chair in Cancer Disparities Research, South Carolina State University, Orangeburg, SC, United States
[g]University of South Carolina, Arnold School of Public Health, Columbia, SC, United States
*Corresponding author: e-mail address: kcartme@clemson.edu

Contents

Abstract

Clinical research is vital to the discovery of new cancer treatments that can enhance health and prolong life for cancer patients, but breakthroughs in cancer treatment are limited by challenges recruiting patients into cancer clinical trials (CT). Only 3–5% of cancer patients in the United States participate in a cancer CT and there are disparities in CT participation by age, race and gender. Strategies such as patient navigation, which is designed to provide patients with education and practical support, may help to

Advances in Cancer Research, Volume 146
ISSN 0065-230X
https://doi.org/10.1016/bs.acr.2020.01.008

overcome challenges of CT recruitment. The current study evaluated an intervention in which lay navigators were utilized to provide patient education and practical support for helping patients overcome barriers to CT participation and related clinical care. A patient barrier checklist was utilized to record patient barriers to CT participation and care, actions taken by navigators to assist patients with these barriers, and whether or not these barriers could be overcome. Forty patients received patient navigation services. The most common barriers faced by navigated patients were fear ($n = 9$), issues communicating with medical personnel ($n = 9$), insurance issues ($n = 8$), transportation difficulties ($n = 6$) and perceptions about providers and treatment ($n = 4$). The most common activities undertaken by navigators were making referrals and contacts on behalf of patients (e.g., support services, family, clinicians; $n = 25$). Navigators also made arrangement for transportation, financial, medication and equipment services for patients ($n = 11$) and proactively navigated patients ($n = 8$). Barriers that were not overcome for two or more patients included insurance issues, lack of temporary housing resources for patients in treatment and assistance with household bills. The wide array of patient barriers to CT participation and navigator assistance documented in this study supports the CT navigator role in facilitating quality care.

1. Introduction

Cancer is the second leading cause of death in the United States (US), second only to heart disease (National Center for Health Statistics, 2017). It is estimated that in, 2017, 600,920 people in the US will die from cancer (American Cancer Society, 2017). Clinical research is vital to the discovery of new cancer treatments that can enhance health and prolong life for cancer patients. However breakthroughs in cancer treatment are limited by challenges recruiting patients into cancer clinical trials (CT). In the US, approximately 20% of cancer patients are eligible for a cancer CT, and 15–25% of eligible patients participate in a CT (National Cancer Institute (NCI). American Society of Clinical Oncology (ASCO), 2010). Thus, only 3–5% of cancer patients in the US actually participate in a cancer CT (National Cancer Institute (NCI). American Society of Clinical Oncology (ASCO), 2010).

Based on epidemiological studies that have examined CT participation enrollment fractions by demographic subgroups, disparities in CT participation among demographic subgroups of the population are evident. The likelihood of CT participation decreases with age. Compared to individuals between the ages of 30–64, those who are ages 65–74 are 0.43 times as likely and those who are ages 75+ are 0.15 times as likely to participate in a CT (Murthy, Krumholz, & Gross, 2004). Men are 1.2–1.8 times as likely as women to participate in cancer CTs (Du, Gadgeel, & Simon, 2006; Murthy et al., 2004). Compared to Caucasians, African Americans (AAs) and Hispanics are approximately 0.7 times as likely to participate in a CT (Murthy et al., 2004). Additionally AAs report more misperceptions and concerns about research than other groups, which may be attributable to historical research mistreatment (Corbie-Smith, Thomas, & St George, 2002; Gadegbeku et al., 2008; Gamble, 1997).

Under-representation of demographic subgroups in research poses a threat to the generalizability of CT results. For example, AAs and the elderly not only have lower rates of CT participation, but also have higher rates of cancer incidence and mortality (Howlader et al., 2011) and poorer overall health status (BRFSS, 2018). Without adequate representation among demographic subgroups in CTs, researchers cannot learn about potential sub-group differences in therapeutic response and toxicities for therapies being tested and ensure the generalizability of CT findings.

Based on averages from observational studies, between 29% and 49% of patients offered a CT option refuse to participate (Jenkins & Fallowfield, 2000; Klabunde, Springer, Butler, White, & Atkins, 1999; Lara et al., 2001). Interventions to provide patients with CT education and practical support to overcome barriers to CT participation and related clinical care are a promising strategy to improve CT participation. Patients considering a CT option face well-documented barriers to CT participation. These include misperceptions and lack of understanding of the CT option and logistical issues such as cost, transportation and complex clinical regimens (Ford et al., 2008). Despite disparities in CT participation among population subgroups, the results of another study that provided structured education for research participation suggests that when AAs and Caucasians are presented with the same research opportunities, they are equally likely to agree to participate (Durant, Legedza, Marcantonio, Freeman, & Landon, 2011). This finding raises the possibility that health disparities in research participation are amenable to intervention.

The National Cancer Institute (NCI) and the American Society of Clinical Oncology (ASCO) summarized consensus recommendations for research strategies that need to be tested for improving patient participation in CTs (Denicoff et al., 2013). Patient navigation has been recommended by the NCI and the ASCO as a strategy that may help to improve recruitment and retention of patients in CTs (Denicoff et al., 2013). In the clinical setting, patient navigation generally refers to strategies that provide personal assistance to help patients overcome specific educational, communication, and logistical barriers to treatment and follow up medical care. Since patient navigation is an individually tailored and interactive intervention, it has strong potential as a strategy to improve personalized CT decision-making and enrollment. To fill this gap in the evidence, our research team carried out a study to evaluate the effect of a lay navigation intervention that utilized lay navigators to help patients overcome the issues that pose barriers to CT participation (Bryant, Williamson, Cartmell, & Jefferson, 2011; Cartmell et al., 2016). The current study reports on the barriers that patients in the study encountered to CT participation and related clinical care, activities taken by navigators to assist over come these barriers, and the extent to which navigators were able to address these barriers.

2. Methods

The purpose of this descriptive quantitative study was to characterize how patient navigators can help cancer patients overcome barriers to cancer CT participation and clinical care. Three specific aims were evaluated to summarize: (1) barriers to CT participation and clinical care experienced among navigated patients; (2) actions taken by navigators to assist patients with these barriers; and (3) the extent to which navigators were able to help patients overcome these barriers. To assess these study aims, navigators used a structured patient barrier checklist to record patient barriers, navigator actions to assist patients, and whether or not they were able to resolve these barriers. This form was based upon the NCI Patient Navigation Research Program's standardized patient log (Freund et al., 2008) and modified to evaluate additional CT-related barriers to care.

2.1 Background of CT navigation intervention

The CT navigation intervention and training program have been previously described in detail (Bryant et al., 2011; Cartmell et al., 2016). Briefly, an overview of the study is provided. The study, which was conducted at three NCI-affiliated cancer centers in the Southeast US, took place between 08/2010 and 10/2011. All study sites had access to the full portfolio of NCI therapeutic cancer trials. To be eligible for the study, patients had to be 18 years of age or older, planning to receive primary therapy at the cancer center and be eligible for a therapeutic CT. The study was approved by the IRBs at each cancer center.

All three study sites had one navigator who worked 20h per week. Patients in the clinic were screened for therapeutic trial eligibility. Potentially eligible CT candidates were invited to enroll in the CT navigation study. The initial navigation started when the CT navigator and the patient watched a 17-min educational video developed by the NCI to provide basic education about CT education. Following the video, the navigator then opened discussion with the patient about the CT, answered their questions about the CT when appropriate, or referred the question to the appropriate clinical team member. The navigator then used the patient barrier checklist (Freund et al., 2008) to screen the patient for barriers to care that might affect clinical treatment and/or CT participation and offered assistance to help mitigate these barriers to care. After this initial session, the navigator assisted the patient with barriers to CT participation and clinical care by: (1) sending reminders to help enhance compliance with the trial protocol, (2) serving as a liaison between patient and clinical team for implementation of the patient's care plan, (3) mitigating logistical barriers to CT participation, and (4) delivering emotional support, either directly by referring them to resources. Navigators contacted patients by phone or personal contact at least once per week to reassess previously identified barriers and identify new barriers.

2.2 Conceptual framework

The Chronic Care Model (CCM) was used in this study to organize the types of roles and responsibilities of navigators that can help patients to overcome barriers to CT participation (Wagner, 1998). This widely used model describes health care system elements to facilitate delivery of high quality chronic disease care. Model elements include: (1) community resources and policies, (2) health system organization, (3) self-management support, (4) delivery system design, (5) decision support and (6) clinical information systems. The CCM has been validated for use across a wide range of chronic diseases and was adapted for use in this navigation intervention. CT navigators support the six elements contained in the CCM. In regard to "community resources and policies," navigators mobilize community resources to meet cancer patient needs by establishing relationships with community services and referring patients for services such as housing and financial assistance. In regard to "organizational commitment to health system changes," the navigation intervention was conceptualized by cancer center leaders to improve low accrual of patients in cancer CTs. This provides evidence of support for the intervention by cancer center leaders. In regard to "self management support," navigators ensure that patients understand their role in maximizing the likelihood of successful treatment.

In regard to "delivery system design," the navigation intervention transforms the process of delivering care by adding a lay navigator to the care team to facilitate more culturally-tailored, patient-centered care. In regard to "decision support," navigators provide patients with evidence-based CT education and facilitate medical information exchange between patient and care team. In regard to "clinical information systems," navigators support the clinical information system by relaying appointment reminders contained in the clinical information system to patients. In the context of the CCM, the navigator's facilitation of these healthcare system elements can promote productive interactions between an informed, active patient and a prepared, proactive practice team, with an end result being better patient care.

2.3 Measurement

A patient barrier checklist was used to document patients' barriers to CT participation and related clinical care and actions taken by navigators to resolve these barriers. This checklist is based on the NCI Patient Navigation Research Program's standardized patient log (Freund et al., 2008) and was adapted to include items specific to CTs. Examples of response options added to incorporate CT specific barriers included patient misperceptions about CTs and patient insurance not covering CT participation. The barrier log included the following variables: date and length of navigation encounters, type of encounter (e.g., home visit, phone call to patient), barriers to CT participation and related clinical care (e.g., transportation, housing, social support) and actions taken by navigators to assist patients with these barriers (e.g., referrals made, education, scheduling appointments). The barrier log enabled collection of the time it took navigators to assist with each type of barrier. These data could be aggregated to track total navigation time to assist with barriers. It was not possible to track overall navigation time because the log did not include a variable to track the time it took to view the CT educational video with the patient in the initial encounter.

Navigators systematically assessed patients in their navigation caseload for barriers that might prevent CT participation or receipt of optimal care using the structured patient barrier checklist. They documented patient barriers and their navigation action plans to resolve these barriers. In addition, a free form text field was added to the barrier checklist to enable the navigators to describe in their own words more about the barriers and their navigation activities. Barriers were initially assessed at entry to navigation and were continually reassessed throughout the CT decision-making and participation period.

2.4 Data analysis

The patient navigator transferred data from the barrier checklist into the REDCap (Research Electronic Data Capture) database for data management. The first step in data analysis was to examine the raw data for patient barriers and navigator actions. Data were examined to determine if there was overlap within any of the barrier categories and if any of the barrier categories could be combined. The two categories of "Perceptions of Treatment" and "Beliefs about Providers" were combined because both of these types of barriers required similar navigation interventions: a need to educate patients regarding beliefs that could jeopardize treatment outcomes.

The frequency of each type of barrier (e.g., transportation) and specific details about the barrier (e.g., public transportation not easily accessible) was calculated. Similarly, the frequency of each type of action taken by navigators to assist patients was calculated overall and by barrier type. Medians and ranges were calculated for: (1) barriers per patient, (2) navigation time per patient and (3) navigation time per barrier. Further assessment described the extent that navigator actions were able to resolve barriers.

3. Results

3.1 Patient characteristics

Table 1 summarizes the characteristics of the 40 patients who received patient navigation services. Most navigated patients were Caucasian (75%), non-Hispanic (98%), male (73%) and married (68%). While most were insured (93%), 69% did not have a college degree and 33% reported being "not at all" or "somewhat" confident to fill out medical forms. Eighty-five percent of participants had non-small cell lung cancer, with the remainder having small cell lung cancer or esophageal cancer. The majority of cancers were late stage cancers, consistent with the expected stage at presentation for lung and esophageal cancers that did not have an evidence-based early detection modality during the study period.

3.2 Patient barriers and navigator actions

The barriers experienced by patients and navigator actions to assist patients with these barriers is summarized in this section.

Table 1 Characteristics of navigated patients.

Characteristic	Category	Navigated patients ($n = 40$)	
		n	%
Age	Mean (SD)	63.10	(9.56)
Race	Caucasian	30	75.0
	African American	10	25.0
Ethnicity	Non-Hispanic	39	97.5
	Hispanic	1	2.5
Gender	Male	29	72.5
	Female	11	27.5
Insurance status	Uninsured	3	7.5
	Insured	37	92.5
Cancer Center Study Site	Charleston, SC	16	40.0
	Savannah, GA	10	25.0
	Spartanburg, SC	14	35.0
Marital status	Not married	13	32.5
	Married/living with partner	27	67.5
Education level	Less than high school	5	12.8
	High school graduate	11	28.2
	Some college/vocational training	11	28.2
Confidence completing Medical forms	Not at all/somewhat	13	33.3
	Quite a bit/extremely	26	66.6
Type of cancer	Non-small cell lung	34	85.0
	Small cell lung	3	7.5
	Esophageal	3	7.5
Stage at diagnosis	Early stage	0	0.0
	Late stage	35	85.0
	Unknown	5	15.0

3.2.1 Patient barriers to care

Fig. 1 provides a summary of the number of barriers documented per patient. Among the 40 patients who participated in the navigation intervention, 18 patients had no documented barriers (45%), 10 patients had 1 barrier (25%), 6 patients had 2 barriers (15%), 2 patients had 3 barriers (5%) and 4 patients had 4–7 barriers (10%). The median number of barriers documented per patient was 1 and ranged from 0 to 7 barriers per patient. The bulk of barriers were clustered in a small subset of navigated patients. Eighty percent of barriers occurred in 30% of navigated patients and 56% of barriers occurred in 15% of patients.

Fig. 2 provides an overview of the type and frequency of these barriers. Fifty barriers were documented among the 22 navigated patients with barriers across the three study sites. The most commonly reported barriers faced

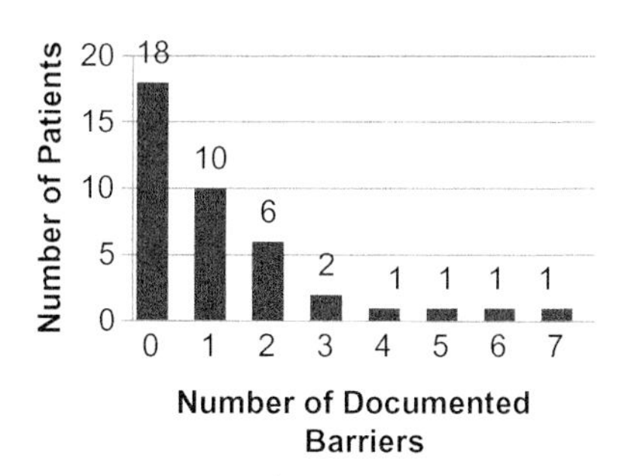

Fig. 1 Number of barriers per navigated patient.

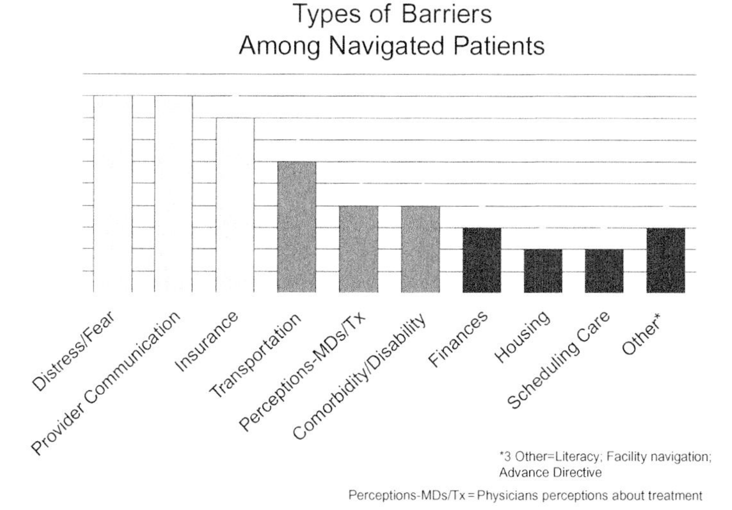

Fig. 2 Types of barriers experienced by navigated patients.

by navigated patients were fear ($n = 9$) and issues regarding communication with medical personnel ($n = 9$), followed by insurance issues ($n = 8$), transportation difficulties ($n = 6$), perceptions about providers and treatment ($n = 4$), disability ($n = 4$), finances ($n = 3$), housing ($n = 2$), scheduling care ($n = 2$) and other barriers. Each of these types of barriers and the actions taken by patients to resolve the barriers are described in further detail.

3.2.2 Navigator activities

Seventy three actions were carried out by navigators to assist the 22 navigated patients who reported barriers to care. These actions are summarized in Table 2. The most common type of activity carried out by navigators was making referrals and contacting patients and other individuals on their behalf with whom they needed to make contact (e.g., support services, family, clinicians; $n = 25$). Navigators also commonly made arrangement for transportation, financial, medication and equipment services for patients ($n = 11$), and spent time proactively navigating patients ($n = 8$).

For the 50 barriers documented by the navigators, the number of contacts and amount of navigation time required per barrier are characterized. The median number of contacts per barrier was 1 and ranged between 1 and 11 contacts. The median reported navigation time per barrier was 20 min and ranged between 17 and 240 min. Navigation time per barrier was less than 30 min for 48% of barriers, between 30 and 59 min for 20% of barriers and between 1 and 4 h for 32% of barriers.

3.2.3 How navigators assisted patients with each type of barrier

Table 3 summarizes the types of barriers experienced by navigated patients and the actions performed by the navigators to assist these patients. The number of barriers does not necessarily equal the number of navigator actions because navigators may have performed no action or many actions per barrier.

3.2.3.1 Fear

The nine fear-related barriers included fear of dying ($n = 4$), needing someone to talk to ($n = 3$), fear that family could not cope with cancer ($n = 1$) and fear of treatment ($n = 1$). To assist with these fears, navigators most frequently provided emotional support and active listening ($n = 5$). They also contacted families to check on them ($n = 1$) and contacted the hospital chaplain to provide additional support for the patient ($n = 1$). For a patient concerned about what would happen to a relative if they died, a navigator educated the patient about available resources ($n = 1$) and educated staff about the patient's individual needs ($n = 1$).

Table 2 Navigator actions to assist patients with barriers ($n = 73$ total actions).

Referrals/direct contact ($n = 25$)	Call patient ($n = 2$)
	Make referral for social services ($n = 2$)
	Make referral for health care services ($n = 1$)
	Directly contact social service agencies ($n = 3$)
	Directly contact family ($n = 9$)
	Directly contact other support system ($n = 4$)
	Directly contact other healthcare providers ($n = 4$)
Arrangements ($n = 11$)	Arrange for transportation ($n = 3$)
	Arrange for financial assistance ($n = 5$)
	Arrange for medication assistance ($n = 2$)
	Arrange for equipment and supplies ($n = 1$)
Proactive navigation (i.e., not related to a barrier) ($n = 8$)	Appointment/info verification/info gathering ($n = 7$)
	Educating staff of patient's special needs ($n = 1$)
Records/recordkeeping ($n = 8$)	Helping patient complete paperwork ($n = 5$)
	Request records/imaging-organize health info ($n = 2$)
	Provide documents to healthcare providers ($n = 1$)
Education ($n = 6$)	Educate verbally ($n = 3$)
	Distribute print and/or audio-visual materials ($n = 3$)
Support ($n = 6$)	Provide emotional support, active listening ($n = 6$)
Accompaniment ($n = 4$)	Accompany to healthcare services ($n = 2$)
	Accompany to other services ($n = 2$)
Scheduling appointments ($n = 3$)	Schedule/reschedule appointment ($n = 1$)
	Call to remind patient of appointment ($n = 2$)
Other ($n = 2$)	Other ($n = 2$)

Table 3 Patient barriers and navigator actions.

Barrier type	Specific barriers	Navigator actions
Fear ($n = 9$)	Fear of dying ($n = 4$)	Directly contact family ($n = 1$) and other support systems ($n = 1$)
	Needs someone to talk to ($n = 3$)	Provide emotional support and active listening ($n = 5$)
	Fear of family not being able to cope ($n = 1$)	Educate patient verbally ($n = 1$) and educate staff of patient's special needs ($n = 1$)
	Fear of treatment ($n = 1$)	Accompany to healthcare services ($n = 1$)
Communication with medical personnel ($n = 9$)	Doesn't understand what asked to do ($n = 3$) regarding when is next appointment, form of medication to take and nutrition needs to remain CT eligible	Directly contact family ($n = 1$) and providers ($n = 3$)
	Medical forms too complicated ($n = 2$)	Appointment verification ($n = 1$)
	Difficulty answering provider questions and incorrectly reporting meds ($n = 2$)	Obtain lab sample from patient ($n = 1$)
	Needs to speak with provider to ask what meds to take for side effects and to schedule appointment ($n = 2$)	Request records/imaging ($n = 1$)
Insurance ($n = 8$)	Can't afford high co-pay/deductible ($n = 1$)	Direct contact with patient ($n = 1$), family ($n = 1$) and support system ($n = 1$)
	Doesn't understand what insurance covers ($n = 1$)	Accompany to other services ($n = 1$); Arrange for financial ($n = 1$) and medication assistance ($n = 2$)
	No health insurance ($n = 1$)	Appointment info/verification ($n = 1$)
	Overwhelmed by insurance paperwork for Medicaid, short term disability, and a cancer policy ($n = 5$)	Helping patient complete paperwork ($n = 2$); Request records/imaging ($n = 1$)
		Provide documents to healthcare providers ($n = 1$)

Transportation ($n = 6$)	Can't afford bus/train fare, gas ($n = 4$)	Direct contact with family ($n = 1$), other support system ($n = 1$) and social service ($n = 1$)
		Arrange Medicaid transportation for patients ($n = 4$) and making referrals to social service agencies (hospital social worker, Department of Social Services [DSS], church) ($n = 4$)
		Helping patient complete paperwork ($n = 2$)
	Public transport not readily available ($n = 2$)	Schedule/reschedule appointment ($n = 1$)
		Appointment/info verification, or info seeking/gathering ($n = 1$)
		Call to remind patient of appointment ($n = 2$)
Perceptions about provider and treatment ($n = 4$)	Felt doctor giving up on them in CT ($n = 2$)	Direct contact with family ($n = 3$), other support system ($n = 1$) and other healthcare providers ($n = 1$)
	Believes treatment medications will make side effects worse ($n = 1$)	Provide emotional support, active listening ($n = 1$)
	Doesn't think CT treatment will help ($n = 1$)	
Medical co-morbidities and disability ($n = 4$)	Physical weakness ($n = 3$)	Directly contact family ($n = 2$) and other healthcare providers ($n = 1$)
		Arrange for equipment and supplies ($n = 1$)
		Accompany to healthcare services ($n = 1$)
	Co-morbidities-obesity ($n = 1$)	Make referral for weight management to get weight under control for treatment ($n = 1$)

Continued

Table 3 Patient barriers and navigator actions.—Cont'd

Barrier type	Specific barriers	Navigator actions
Finances ($n = 3$)	Can not afford household bills ($n = 2$)	Directly contact social service ($n = 1$) and other support system resources such as American Cancer Society (ACS), Wal-Mart, and utility companies ($n = 1$)
		Arrange for financial assistance ($n = 1$)
	Can't afford over the counter supplements needed during treatment ($n = 1$)	Arrange for financial assistance ($n = 1$) via product coupons and website to help with cost of ancillary medical products
Housing ($n = 2$)	Need temporary housing for treatment ($n = 2$)	Educate verbally about temporary housing options ($n = 1$)
		Distribute print/audio-visual materials (hotel list) ($n = 1$)
		Appointment/info verification, or info seeking/gathering for housing ($n = 1$)
Problem scheduling care ($n = 2$)	Appointment date/time not convenient ($n = 2$)	Appointment/info verification, or info seeking/gathering ($n = 2$)
Other ($n = 2$)	Navigation of healthcare facility ($n = 1$)	Accompany to other services ($n = 1$)
	Advance directive ($n = 1$)	Referral to provider ($n = 1$)
Literacy ($n = 1$)	Unfamiliar/do not understand medical terms ($n = 1$)	Helping patient complete medical paperwork ($n = 1$)

Quotations from the navigators illustrate how navigators assisted patients with issues of fear. One navigator described helping a patient who was afraid of dying and concerned that the medication would not help: "I called the chaplain in to talk with the patient...and also asked the patient to discuss this concern with the doctor when he came in." Another navigator described providing support for a frightened family member: "The patient came in with his wife...he was in a wheel chair and very frail. When I entered the room, I found that his wife was shaken. I sat with the patient and his wife. I simply tried to comfort her, giving her a hug, and letting her know that all members of the team care about them and are here to assist them. I know at times caregivers can feel like they are alone and have no one to turn to." For another patient who feared treatment, a navigator described: "I sat in the treatment room with her for two hours talking to her and calming her down because her husband had to leave."

3.2.3.2 Communication with providers

The nine barriers related to communication with medical providers included patients not understanding what the clinical team had asked them to do ($n = 3$), complicated medical forms ($n = 2$) and other communication barriers ($n = 4$). For patients who did not understand what clinicians had asked them to do, patients needed to know when clinic visits were scheduled ($n = 1$), what form of a prescribed medication to take ($n = 1$) and how to follow provider instructions for modifying diet to meet CT eligibility requirements ($n = 1$). A single patient encountered two barriers when medical forms were too complicated. In trying to get disability paperwork completed, assistance was needed to obtain disability forms and to get the provider to complete the forms. The other communication barriers occurred when patients had difficulty communicating medical history and current medications to providers ($n = 2$) and when patients needed help setting up return visits for routine appointments and for re-drawing lab work for CT eligibility ($n = 2$).

Quotations from the navigators illustrate how navigators facilitated patient-provider communication. Navigators frequently described how they clarified medication questions for patients. One navigator said: "In talking to the patient, I found out that the patient's wife was not giving him all the prescribed medications and that she had not provided a complete list of all the medications, including those from the patient's primary care provider." The navigator then arranged for the patient and provider to meet to clarify the medication regimen. Other navigators described helping with medication communication by: "speaking with the provider about medical questions the patient had related to side effects and medications that could be taken" and "making a journal for the patient to track their medications." Another navigator described how she facilitated communication with providers about complex medical paperwork. She obtained disability paperwork, sent it to the provider to complete and "went to check on the disability letter for the patient so that they did not have to make a special trip to the hospital for it." Another navigator described helping a patient understand and comply with lab work needed to establish CT eligibility: "The patient was given a urine container for her specimen and a doctor's order. The patient will have to return with the 24 h urine sample for creatinine clearance. The patient's appointment was rescheduled so she could bring back the urine." The navigator ensured that the patient understood when to return to clinic with the specimen.

3.2.3.3 Insurance

The eight insurance-related barriers included being overwhelmed by insurance paperwork ($n=5$), being unable to afford a medication co-pay ($n=1$), not understanding what insurance would cover ($n=1$) and not having insurance ($n=1$). Three of the five barriers related to insurance paperwork occurred for one patient. These barriers included needing help getting medical records for disability insurance, needing physician prescription for a medication and needing help with paperwork required by insurance to cover medication. The other two insurance paperwork-related barriers occurred in one patient who needed help with paperwork to activate a cancer insurance policy and another patient whose provider had not completed their disability insurance paperwork correctly.

For patients with insurance barriers, navigators most commonly helped arrange financial ($n=1$) and medication assistance ($n=2$). Navigators also completed insurance paperwork ($n=2$), obtained records and imaging needed by insurance companies ($n=1$) and provided insurance documents to healthcare providers. Other navigator actions to assist with insurance barriers included making contact with patients ($n=1$), their families ($n=1$) and support systems ($n=1$), accompanying patients to other service providers ($n=1$) and verifying patient appointments ($n=1$).

Quotations from the navigators further illustrate how navigators assisted patients with insurance barriers. One navigator described: "I helped the patient fill out papers for a cancer policy they just had taken out. I faxed papers in for her and gave her a copy of the confirmation that it went through. Another navigator said: "I helped a family member get medical records together so she could go apply for his Medicaid and disability check." Other navigators helped patients with insurance paperwork by referring them to a social worker or financial counselor. One navigator described how she helped a patient to sort out what their insurance would cover: "I called CVS and asked why Medicaid was not paying for the medicine and she reported that it was showing inactive. I called Medicaid the next morning and she explained that he did not have regular Medicaid but it pays his Medicare premium. I explained the situation to the doctor and his nurse, and the nurse said she would work on getting him assistance."

3.2.3.4 Transportation

The six transportation-related barriers included being unable to afford bus/train fare or gas ($n = 4$) and public transportation not being easily accessible ($n = 2$). Four of the six barriers were for one patient who needed help with multiple trips to the clinic for treatment appointments. The other two barriers occurred in two patients who also needed help with transportation to the clinic for treatment appointments.

For patients with transportation barriers, navigators assisted with transportation by (1) arranging Medicaid transportation for Medicaid patients ($n = 4$) and (2) referring patients without Medicaid to resources for assistance with gas and volunteers to provide transportation to appointments ($n = 4$). For patients without Medicaid, navigators made referrals to hospital social workers, the department of social services and churches for transportation assistance and helped to complete financial paperwork for these services ($n = 2$). Navigators spent time scheduling ($n = 1$), verifying ($n = 1$) and sending reminders about transportation appointments ($n = 2$). Navigators also spent time contacting patients' family ($n = 1$), other support system resources ($n = 1$) and social service providers ($n = 1$) regarding transportation.

Quotations from navigators further illustrate how navigators helped patients with transportation barriers. Navigators described: "I made reservations with Logisticare for the patient to be picked up at home and transported to the facility" and "I made arrangements for the patient to be transported by Logisticare for their Gamma Knife procedure." Navigators also described how they helped patients not eligible for Medicaid transportation: "I referred the patient to the social worker to help with the resources needed to get to the clinic for treatment on trial" and "I called the social worker and told him this gentleman needed help getting transportation to and from the clinic for treatment."

3.2.3.5 Patient perceptions about providers and treatment

The four barriers related to patient perceptions about providers and treatment including feeling the provider was giving up on their treatment ($n = 2$), a family member hesitant to administer medication due to unwanted side effects ($n = 1$) and a belief that treatment would not help ($n = 1$). To assist with these barriers, navigators contacted family ($n = 3$), healthcare providers ($n = 1$) and other support systems ($n = 1$) to help talk through patient concerns. Navigators also provided emotional support and active listening ($n = 1$).

One navigator described how she assisted a patient and his wife who felt their doctor had given up on them when he took the patient off of the CT due to toxicities experienced during treatment. The patient expressed he had hoped he was going to get a 'better chance' at getting to complete the treatment regimen he was taking in the CT. I assured the patient and his wife that this decision was done because of his health, and that they should bring up their disappointment with their doctor. The couple brought up their feelings with the doctor, and were told that when the patient was able to do housework, he would be treated again, until then, he was not healthy enough to do the treatments." Another navigator described: "I talked with the patient about why they felt they might want another doctor. The patient's wife just wanted to make sure that she was getting the best care for her husband and that the doctor was not just giving up on him. I reassured the patient and his family that the doctor was a good doctor and that he treats all of his patients with care no matter what stage of cancer they might have."

One navigator described how she helped to ensure that a patient was taking medications correctly: "After a series of phone calls, I found out that the patient was not getting his medications as directed. The patient's wife stated that she read the side effects on the package, one of which was nausea and vomiting. This is one of the symptoms the patient was experiencing. In an effort to keep the patient from experiencing the side effects, she stopped giving him the medication. When the patient and his wife shared this with me, I informed the doctor and he conducted a teaching session to inform the couple how to take each medication and what each one did. The doctor asked me to retype the list of medications and print in large letters for the patient's wife to use as a guide. He also asked me to check up on the patient to ensure they were taking the medication appropriately. The patient expressed their understanding afterwards." Another navigator described how she assisted a patient who was afraid that participation in a CT might worsen her health. The navigator described: "I called in the nurse and had her to speak with the patient and his family again about the clinical trial and she provided them with some more education about the trial."

3.2.3.6 Medical co-morbidities and disability

Four barriers related to medical co-morbidities occurred in patients whose poor physical health was making it difficult to come for treatment ($n = 4$). To assist with issues related to poor physical health, navigators spent time contacting family members ($n = 2$) and healthcare providers ($n = 1$). Navigators also arranged for a wheelchair ($n = 1$), accompanied patients to a healthcare appointment ($n = 1$) and made a referral to a nutritionist ($n = 1$).

Quotations from navigators further illustrate how navigators assisted patients with medical co-morbidities and disability. One navigator described: "The doctor wants to take the patient off of protocol because he is losing weight rapidly and has become very weak. I read the patient's office notes and went and discussed the patient's weight with the nurse. I told her that I would refer the patient to the nutritionist when I saw the patient. I went to see the patient and suggested him seeing the nutritionist and the sister agreed. I went to the nutritionist's office and asked if she could see the patient while he was in infusion. Another navigator described: "The patient's wife expressed the desire to get a handicapped parking placard. I provided the patient's wife with the application they needed and informed the nurse they would need a prescription when they returned for their appointment." Another navigator said: "I sat with the patient and his wife, and she expressed that because of her husband's weakness, she is afraid to leave him at home alone, and as a result has been neglecting her doctor's appointments and other errands. I suggested hospice respite care, and she said that she would really like that because she would prefer to have medical staff available to her husband in her absence. I talked to the doctor and nurse to make sure this process was started in a timely fashion."

3.2.3.7 Finances

Three finance-related barriers included being unable to afford household bills ($n = 2$) and an over-the-counter skin care treatment product recommended by their provider ($n = 1$). To assist patients with financial barriers, navigators contacted social service ($n = 1$) and other support system resources such as the American Cancer Society (ACS), Wal-Mart, and utility companies ($n = 1$) for assistance. They also arranged for financial assistance when resources were available ($n = 2$).

Quotations from navigators illustrate how navigators assisted patients with financial barriers. One navigator described her attempts to locate a source of funding for a patient's prepaid cell phone service: "I contacted the American Cancer Society, local television stations, Wal-Mart, and the company that provided the service. I was unable to get any of these places to offer a donation. I was able to secure a private donation from an employee without soliciting. She heard the issue and decided to help out." Another navigator described her attempt to obtain financial assistance for help with household bills: "I called the social worker so that he could come and speak with her. The social worker called and talked with her." Another navigator described helping a patient pay for a treatment supplement: "The patient expressed the expense related to Aveeno products which were recommended by their provider. I provided the patient with coupons for Aveeno products and a website to help with the cost of ancillary medical products."

3.2.3.8 Housing

Two housing-related barriers occurred in patients who needed help with temporary local housing during cancer treatment. Navigators attempted to assist patients with housing barriers by gathering information about available housing resources ($n = 1$), educating patients about the availability of housing resources ($n = 1$) and providing a list of hotels that provide discounted rates for patients in treatment ($n = 1$). One navigator described: "The patient requested information about staying in Hope Lodge on those days that they would be here for treatment. After some investigation, I informed the patient that because they would only need to be here once a week, the Hope Lodge would not be available to them." Another navigator described: "The patient and family wanted to make lodging accommodations because they would be in town for three days straight, and there was a waiting list at the Hope Lodge. I provided them with a list of hotels in the area that provide discounted rates." Barrier resolution was defined in this study as completing actions to assist patients with a barrier, regardless of whether the barrier was overcome. These barriers related to housing highlight that sometimes barriers were resolved by navigators in sub-optimal ways from the patient perspective.

3.2.3.9 System of scheduling care

Two system-related barriers occurred in patients who were given inconvenient appointment times. To assist patients with these barriers, navigators sought information from appointment schedulers to try to change the appointment times ($n = 2$). One navigator described: "The patient needed all appointments to be rescheduled for the afternoon because it is hard for him to get up in the morning so I got that taken care of." Another navigator said: "The patient's wife wanted to change appointment dates because they are scheduled to come in on 2 days back to back. Because they live 5 h away, this will be a long drive. However, the appointment could not be changed because of the nature of the first visit and cannot be done while the patient receives chemotherapy."

3.2.3.10 Other barriers

Other barriers included a literacy issue in which patients did not understand medical terminology ($n = 1$), had difficulty navigating the healthcare facility ($n = 1$) and needed help with an advance directive ($n = 1$). A navigator described helping a patient with literacy issue by assisting him to complete paperwork: "I read the surveys out for him and he answered the questions. He completed the 9th grade but does not understand all the medical stuff that the doctors are saying." Another navigator described helping a patient navigate the hospital by accompanying them to where they needed to go: "The patient and husband were very hungry, but did not feel they could navigate efficiently to the cafeteria and back. I accompanied them to the cafeteria, and afterwards to their car." A navigator assisted a patient with an advance directive by "referring the patient to the chaplain to help the family fill out the advance directive."

3.2.3.11 Barriers not identified

No barriers were recorded for the following barrier types listed on the barrier checklist: citizenship concerns, need for language/interpreter, childcare Issues, family/community issues, out of town/country, social/practical support and work schedule conflict.

3.3 Resolution of barriers

For each of the 50 barriers encountered by patients, navigators were asked if they were able to resolve these barriers. Out of the 50 patient barriers, navigators documented that 44 of these barriers were resolved (88%), 4 were not resolved (8%) and did not provide information about barrier resolution for 2 barriers (4%). Three of the four unresolved barriers related to insurance. For a patient who did not know what insurance would cover, the navigator found out that Medicaid would not cover medication, but this ascertainment did not result in the patient obtaining medication coverage. Navigators documented that barriers were not resolved for a patient who needed help with medication copays and a patient whose doctor had filled out short term disability paperwork incorrectly. The fourth documented unresolved barrier was related to a patient's fear of dying. While the navigator provided emotional support, she did not consider that this support was sufficient to overcome the patient's fear of dying.

The two barriers for which there was no documentation of whether the barrier was resolved related to communication with medical personnel. One of these barriers was for a patient who needed help bringing in a urine test to retest lab value needed for CT eligibility. The other barrier was for a patient who was having difficulty answering the physician's questions about their medical history and medications. No additional information is available for these two patients.

While navigators documented that housing barriers were resolved, their notes suggested a lack of availability of robust housing assistance for patients during cancer treatment. For both patients who needed financial assistance with short term housing, neither received housing assistance due to not being in town long enough to be eligible for housing assistance or due to the housing resource being completely booked. Hotel discounts were available for patients which may provide some assistance to patients who need housing during cancer treatment.

3.4 Patient barriers and navigator actions to address barriers

Findings from this study related to patient barriers to CT participation and the actions of navigators to help patients overcome these barriers are discussed below. First, the barriers experienced by patients are discussed, followed by evidence for the navigator's role in helping patients to overcome these barriers. Next, the study findings related to the intensity of navigation services are discussed. Finally, recommendations are made for coordinating the navigator role in the cancer clinic setting based upon our study findings.

3.4.1 Patient barriers identified

The three most common barriers observed in the current study were: (1) fear about cancer, (2) difficulty communicating with providers and (3) insurance issues. Results from our study were compared with those from two navigation studies conducted among breast and colorectal cancer patients (Carroll, Winters, Purnell, Devine, & Fiscella, 2011; Hendren et al., 2011).

Unlike our study, fear was not documented as a major barrier in either of these studies. Consistent with our study results, one of these two comparison studies documented provider communication difficulties as one of the top three barriers to care (Carroll et al., 2011; Hendren et al., 2011) and both of these studies documented insurance issues as one of the top three barriers (Carroll et al., 2011; Hendren et al., 2011). Lack of transportation ranked as the third leading barrier in one of these studies (Carroll et al., 2011) and as the fourth leading barrier in our study. Unlike our study, social support was documented as being one of the top three barriers to care in both of these comparison studies (Carroll et al., 2011; Hendren et al., 2011).

As patient navigation expands into cancer treatment, CTs and post-treatment, it is important to understand the navigator role. Because we tested navigation in a CT setting with late stage cancer patients, we wanted to identify patient barriers in this population. We observed several unique cancer barrier burdens to this population that are not commonly reported in the navigation literature. First, patients commonly suffered from fear/emotional distress and physical symptoms related to late stage disease. Second, patients commonly encountered barriers transitioning from work-related to disability-related income/insurance. Third, patients needed help to coordinate additional tests required for CT eligibility.

The finding of high emotional distress and physical symptoms provides insight into navigation needs for late stage cancer patients. For example, fear was the most frequently reported barrier, some patients were upset their doctor had given up on their CT participation and some patients had difficulty balancing medication compliance with distressing medication side effects. Distress was not cited as a major barrier in two navigation studies of breast and colorectal cancer patients with earlier stage, more curable cancers (Carroll et al., 2011; Hendren et al., 2010). Our study finding that patients need help with symptom distress is supported by the call for formal symptom distress services in cancer care, as described in the landmark Institute of Medicine Report "From Cancer Patient to Cancer Survivor: Lost in Transition" (Committee on Cancer Survivorship: Improving Care and Quality of Life, 2005).

The finding that patients needed intensive support to transition from employer-based insurance to other types of insurance such as short and long term work disability, Consolidated Omnibus Budget Reconciliation Act (COBRA) coverage, Medicaid and Medicare highlights patient difficulties resulting from a fragmented healthcare system. Patients, most of whom started cancer treatment with employer-based insurance, often had to rapidly search for alternative sources of insurance when faced with a disabling cancer diagnosis. The tasks of helping patients fill out paperwork, get medical documentation to support disability insurance/income and appeal denials were some of the more complex and time-consuming services provided by navigators.

4. Discussion

The wide array of patient barriers to CT participation and navigator assistance documented in this study supports the CT navigator role in facilitating quality care as described in the CCM. Navigators' work was roughly evenly distributed across three types of tasks. First navigators provided case management to enhance treatment compliance. Specifically, navigators facilitating patient adherence to medication regimens, scheduled care visits and additional tests required to establish CT eligibility. As described in the CCM, provision of case management to support treatment compliance is essential for ensuring quality of patient care. Second navigators facilitated emotional support for patients via active listening and referral for emotional support. As described in the CCM, identifying and addressing patients' emotional needs that could impede cancer treatment is essential to prepare patients for their self-management role in treatment. Third navigators mobilized community and cancer center resources to help patients overcome logistical barriers to care as recommended in the CCM. In this role, navigators linked patients with resources, particularly related to financial assistance, disability determination and insurance coverage.

Ultimately, we wanted to ascertain if navigators were able to assist patients to overcome barriers to CT enrollment and completion. Navigators reported being able to resolve 88% of identified barriers. However, resolution of barriers was defined by the navigation team as having completed the work process for that barrier, not necessarily that the patient's barrier was overcome. Examples are provided to illustrate this concept. A navigator documented that a housing barrier was resolved by providing a patient with a list of discounted hotels. Another navigator documented that an insurance barrier was resolved by confirming that the patient's insurance would not cover treatment. Based on how "resolution of barrier" was defined, the navigators accurately recorded the barrier resolution outcome. In future navigation interventions, it will be important to track both whether the patient's barrier was resolved and to what extent the navigator was able to garner resources and support to assist the patient.

We were ultimately able to assess whether patient barriers were overcome by relying on text data entered by navigators describing their work to assist each patient. Barriers that were not overcome for two or more patients included: (1) insurance issues, (2) lack of temporary housing resources for patients in treatment and (3) assistance with household bills. These data provide valuable program planning data for assessment of unmet patient resource needs.

To inform the staffing needs for future CT navigation interventions, we examined the intensity of navigation services. The number of barriers per patient and time required to navigate each patient varied dramatically. About half of patients had no specific barriers to care, with 10% having 4–7 barriers. The overall median time required to navigate each patient's barriers was 80 min and ranged from 10 min to 350 min per patient. Unfortunately, the barrier log did not include a variable to enter the time required for the initial patient visit to: (1) view the NCI educational video with the patient, (2) assess and address their questions about CTs and (3) conduct an initial barrier assessment with the patient. Most likely though, this visit would take between 20 and 50 min, depending on the amount of interaction between patient and navigator. This extra time would need to be taken into account to estimate overall navigation time.

The finding of varied service intensity among navigated patients supports the idea of offering multiple service levels for CT navigation. Some navigation studies conducted in patients during cancer screening and treatment have implemented "light," "moderate," and "intensive" navigation (Ell, Vourlekis, Lee, & Xie, 2007; Ell et al., 2009). This tiered-approach allows program managers to systematically vary the intensity of the navigation program protocol (e.g., contact intervals, duration of navigation consults, type of navigator (lay, social worker, nurse)) according to patient needs. This potential navigation program refinement would support one of the recommendations in the CCM to organize program systems to facilitate efficient and effective care. By building navigation service levels into the IT system, patients with complex service needs could be triaged for more intensive navigation services (Wagner, 1998).

With respect to the intensity of navigation services compared with other CT navigation interventions, in one CT navigation study navigators assisted a rural, medically underserved American Indian population to facilitate cancer treatment and CT participation. A median initial visit time of 40 min and follow up visit time of 15 min per visit was reported, with a median of 12 visits per patient (Guadagnolo et al., 2011). Compared to our average barrier navigation time of 90 min plus an extra 20–50 min per patient for viewing and discussing the CT navigation video, these comparison data suggest that the navigation services provided in the study by Guadagnolo et al. were more intensive than in the present study.

There are several possible explanations for why our service intensity was lower than in this comparison study. First, navigated patients in our study may have experienced fewer barriers to care and required less navigation time. This is possible because our navigation program served all patients and not just those identified as poor or medically underserved. Second, our navigators may not have adequately reassessed ongoing barriers and identified new barriers routinely. Navigators were trained and regularly reminded to contact patients at least weekly, but there was no process to monitor that navigators made contact with patients each week. Third, navigators may not have recorded all their activities in the navigation program database. To record navigation encounters and time spent with patients, navigators had to go into each patient record to enter these details each time they performed an activity. Fourth, receipt of intensive ongoing treatment in a multidisciplinary team setting may automatically address issues that in other settings would generate navigator action. The low documented number of barriers and navigation time per patient in our study highlights the need to rigorously: (1) document the intensity of CT navigation services using the barrier checklist and (2) monitor the fidelity in which barriers and navigator actions are recorded in future CT navigation research.

Based on the broad range of patient barriers observed in this pilot study, careful coordination of work activities between the CT navigator and their clinical partners such as clinic nurses, social workers and financial counselors can help to ensure coordination of patient care and avoid duplication of efforts. In fact, one of the recommendations of the CCM is to design tasks and distribute work among team members to ensure efficient delivery of patient care. We identified a number of patient needs that could be addressed by the navigator or by referral to a team member such as a social worker, financial counselor or nurse. For example, a navigator could address a patient's fear of cancer by active listening or referral to a social worker. Similarly, a navigator could assist with disability paperwork or refer the patient to a financial counselor. Decisions about whether navigators should directly assist patients or refer patients to other service providers may vary by factors such as type of barrier, navigator skills and attributes, patient preferences and available resources. Nonetheless, the potential for overlap between the role of navigators and other clinical team members suggests that careful integration of the navigator role and responsibilities within the clinical team is critical.

In summary, approximately 50% of navigated patients had one or more specific barriers to care. Most barriers were concentrated in a relatively small proportion of patients. This distribution of patient barriers demonstrates that there were "heavy users" and "non-users" of navigation. CT navigators assisted patients with a wide range of barriers related to communication with the clinical team, practical barriers to care and distress.

Acknowledgment

This project was funded by the National Cancer Institute through grant #P30-CA13831302.

References

American Cancer Society. (2017). *Cancer facts & figures 2017*. Atlanta, GA, accessed 12/23/19 at https://www.cancer.org/research/cancer-facts-statistics/all-cancer-facts-figures/cancer-facts-figures-2017.html.

BRFSS, & Centers for Disease Control and Prevention (CDC). (2018). *Behavioral risk factor surveillance system survey data*. Atlanta, Georgia: U.S. Department of Health and Human Services, Centers for Disease Control and Prevention.

Bryant, D. C., Williamson, D., Cartmell, K., & Jefferson, M. (2011). A lay patient navigation training curriculum targeting disparities in cancer clinical trials. *Journal of National Black Nurses' Association*, *22*(2), 68–75. Retrieved from https://www.ncbi.nlm.nih.gov/pubmed/23061182.

Carroll, J. K., Winters, P. C., Purnell, J. Q., Devine, K., & Fiscella, K. (2011). Do navigators' estimates of navigation intensity predict navigation time for cancer care? *Journal of Cancer Education*, *26*(4), 761–766. https://doi.org/10.1007/s13187-011-0234-y.

Cartmell, K. B., Bonilha, H. S., Matson, T., Bryant, D. C., Zapka, J. G., Bentz, T. A., et al. (2016). Patient participation in cancer clinical trials: A pilot test of lay navigation. *Contemporary Clinical Trials Communications*, *3*, 86–93. https://doi.org/10.1016/j.conctc.2016.04.005.

Committee on Cancer Survivorship: Improving Care and Quality of Life. (2005). In M. Hewitt, S. Greenfield, & E. Stovall (Eds.), *From cancer patient to cancer survivor: Lost in transition*, (p. 2). Washington, DC: The National Academies Press.

Corbie-Smith, G., Thomas, S. B., & St George, D. M. (2002). Distrust, race, and research. *Archives of Internal Medicine*, *162*(21), 2458–2463. Retrieved from http://www.ncbi.nlm.nih.gov/pubmed/12437405.

Denicoff, A. M., McCaskill-Stevens, W., Grubbs, S. S., Bruinooge, S. S., Comis, R. L., Devine, P., et al. (2013). The National Cancer Institute-American Society of clinical oncology cancer trial accrual symposium: Summary and recommendations. *Journal of Oncology Practice*, *9*(6), 267–276. https://doi.org/10.1200/JOP.2013.001119.

Du, W., Gadgeel, S. M., & Simon, M. S. (2006). Predictors of enrollment in lung cancer clinical trials. *Cancer*, *106*(2), 420–425. https://doi.org/10.1002/cncr.21638.

Durant, R. W., Legedza, A. T., Marcantonio, E. R., Freeman, M. B., & Landon, B. E. (2011). Willingness to participate in clinical trials among African Americans and whites previously exposed to clinical research. *Journal of Cultural Diversity*, *18*(1), 8–19. Retrieved from http://www.ncbi.nlm.nih.gov/pubmed/21526582.

Ell, K., Vourlekis, B., Lee, P. J., & Xie, B. (2007). Patient navigation and case management following an abnormal mammogram: A randomized clinical trial. *Preventive Medicine*, *44*(1), 26–33. https://doi.org/10.1016/j.ypmed.2006.08.001.

Ell, K., Vourlekis, B., Xie, B., Nedjat-Haiem, F. R., Lee, P. J., Muderspach, L., et al. (2009). Cancer treatment adherence among low-income women with breast or gynecologic cancer: A randomized controlled trial of patient navigation. *Cancer*, *115*(19), 4606–4615. https://doi.org/10.1002/cncr.24500.

Ford, J. G., Howerton, M. W., Lai, G. Y., Gary, T. L., Bolen, S., Gibbons, M. C., et al. (2008). Barriers to recruiting underrepresented populations to cancer clinical trials: A systematic review. *Cancer*, *112*(2), 228–242. https://doi.org/10.1002/cncr.23157.

Freund, K. M., Battaglia, T. A., Calhoun, E., Dudley, D. J., Fiscella, K., Paskett, E., et al. (2008). National Cancer Institute Patient Navigation Research Program: Methods, protocol, and measures. *Cancer*, *113*(12), 3391–3399. https://doi.org/10.1002/cncr.23960.

Gadegbeku, C. A., Stillman, P. K., Huffman, M. D., Jackson, J. S., Kusek, J. W., & Jamerson, K. A. (2008). Factors associated with enrollment of African Americans into a clinical trial: Results from the African American study of kidney disease and hypertension. *Contemporary Clinical Trials*, *29*(6), 837–842. https://doi.org/10.1016/j.cct.2008.06.001.

Gamble, V. N. (1997). Under the shadow of Tuskegee: African Americans and health care. *American Journal of Public Health*, *87*(11), 1773–1778. Retrieved from http://www.ncbi.nlm.nih.gov/pubmed/9366634.

Guadagnolo, B. A., Boylan, A., Sargent, M., Koop, D., Brunette, D., Kanekar, S., et al. (2011). Patient navigation for American Indians undergoing cancer treatment: Utilization and impact on care delivery in a regional healthcare center. *Cancer*, *117*(12), 2754–2761. https://doi.org/10.1002/cncr.25823.

Hendren, S., Chin, N., Fisher, S., Winters, P., Griggs, J., Mohile, S., et al. (2011). Patients' barriers to receipt of cancer care, and factors associated with needing more assistance from a patient navigator. *Journal of the National Medical Association*, *103*(8), 701–710. Retrieved from http://www.ncbi.nlm.nih.gov/pubmed/22046847.

Hendren, S., Griggs, J. J., Epstein, R. M., Humiston, S., Rousseau, S., Jean-Pierre, P., et al. (2010). Study protocol: A randomized controlled trial of patient navigation-activation to reduce cancer health disparities. *BMC Cancer*, *10*, 551.

Howlader, N., Noone, A. M., Krapcho, M., Neyman, N., Aminou, R., Altekruse, S. F., et al. (2011). *SEER cancer statistics review, 1975–2009 (vintage 2009 populations)*. Bethesda, MD: National Cancer Institute.

Jenkins, V., & Fallowfield, L. (2000). Reasons for accepting or declining to participate in randomized clinical trials for cancer therapy. *British Journal of Cancer*, *82*(11), 1783–1788. https://doi.org/10.1054/bjoc.2000.1142.

Klabunde, C. N., Springer, B. C., Butler, B., White, M. S., & Atkins, J. (1999). Factors influencing enrollment in clinical trials for cancer treatment. *Southern Medical Journal*, *92*(12), 1189–1193. Retrieved from http://www.ncbi.nlm.nih.gov/pubmed/10624912.

Lara, P. N., Jr., Higdon, R., Lim, N., Kwan, K., Tanaka, M., Lau, D. H., et al. (2001). Prospective evaluation of cancer clinical trial accrual patterns: Identifying potential barriers to enrollment. *Journal of Clinical Oncology*, *19*(6), 1728–1733.

Murthy, V. H., Krumholz, H. M., & Gross, C. P. (2004). Participation in cancer clinical trials: Race-, sex-, and age-based disparities. *JAMA*, *291*(22), 2720–2726. https://doi.org/10.1001/jama.291.22.2720.

National Cancer Institute (NCI), & American Society of Clinical Oncology (ASCO). (2010). *Cancer trials accrual symposium: Science and solutions*. .

National Center for Health Statistics. (2017). *Health, United States, 2016: With chartbook on long-term trends in health*. Hyattsville, MD: National Center for Health Statistics.

Wagner, E. H. (1998). Chronic disease management: What will it take to improve care for chronic illness? *Effective Clinical Practice*, *1*(1), 2–4. Retrieved from http://www.ncbi.nlm.nih.gov/pubmed/10345255.

CHAPTER EIGHT

Project PLACE: Enhancing community and academic partnerships to describe and address health disparities

Nadine J. Barrett[a,b,c,*], Kearston L. Ingraham[a], Kenisha Bethea[c], Pao Hwa-Lin[d,e], Maritza Chirinos[f], Laura J. Fish[a], Schenita Randolph[g], Ping Zhang[h], Peter Le[i], Demetrius Harvey[j,k], Ronald L. Godbee[l], Steven R. Patierno[m]

[a]Duke Cancer Institute, Duke University Medical Center, Durham, NC, United States
[b]Department of Family Medicine and Community Health, Duke School of Medicine, Durham, NC, United States
[c]Duke Clinical and Translational Science Institute, Duke University School of Medicine, Durham, NC, United States
[d]Chinese Christian Church, Raleigh, NC, United States
[e]Department of Medicine, Duke University School of Medicine, Durham, NC, United States
[f]El Centro Hispano, Durham, NC, United States
[g]Duke University School of Nursing, Durham, NC, United States
[h]Chinese American Friendly Association, Raleigh, NC, United States
[i]St. Joseph's Primary Care, Raleigh, NC, United States
[j]Black Men's Health Initiative, Wilson, NC, United States
[k]Alumni Chapter of Kappa Alpha Psi Fraternity, Inc., Smithfield, NC, United States
[l]The River Church, Durham, NC, United States
[m]Department of Medicine, Division of Medical Oncology, Duke University Medical Center, Durham, NC, United States
*Corresponding author: e-mail address: nadine.barrett@duke.edu

Contents

Advances in Cancer Research, Volume 146
ISSN 0065-230X
https://doi.org/10.1016/bs.acr.2020.01.009

Abstract

Achieving cancer health equity is a national imperative. Cancer is the second leading cause of death in the United States and in North Carolina (NC), where the disease disproportionately impacts traditionally underrepresented race and ethnic groups, those who live in rural communities, the impoverished, and medically disenfranchised and/or health-disparate populations at high-risk for cancer. These populations have worse cancer outcomes and are less likely to be participants in clinical research and trials. It is critical for cancer centers and other academic health centers to understand the factors that contribute to poor cancer outcomes, the extent to which they impact the cancer burden, and develop effective interventions to address them. Key to this process is engaging diverse stakeholders in the development and execution of community and population health assessments, and the subsequent programs and interventions designed to address the need across the catchment area. This chapter describes the processes and lessons learned of the Duke Cancer Institute's (DCI) long standing community partnerships that led to Project PLACE (Population Level Approaches to Cancer Elimination), a National Cancer Institute (NCI)-funded community health assessment reaching 2315 respondents in 7 months, resulting in a community partnered research agenda to advance cancer equity within the DCI catchment area. We illustrate the application of a community partnered health assessment and offer examples of strategic opportunities, successes, lessons learned, and implications for practice.

1. Introduction

Achieving cancer health equity is a national imperative. Cancer is the second leading cause of death in the United States (US) and the

National Cancer Institute (NCI) predicts an estimated 1,762,450 new cancer cases and approximately 606,880 cancer deaths in the United States in 2019 (SEER, 2019). Likewise, cancer is the leading cause of death with estimates of almost 60,000 new cases and 20,410 deaths in North Carolina (NC), which has a significant number of counties and communities comprised of a high proportion of racially and ethnically diverse, rural, impoverished, medically disenfranchised and/or health-disparate populations at high-risk for cancer with limited access to resources and services (SEER, 2019). These populations have worse cancer outcomes and are less likely to be participants in clinical research and trials. For effective cancer control, it is critical for cancer centers and other academic health centers to understand the factors that contribute to poor cancer outcomes and the extent to which they impact the cancer burden and develop effective interventions to address them (Paskett & Hiatt, 2018). The key to this process is engaging diverse stakeholders in the development and execution of community and population health assessments, and the subsequent programs and interventions designed to address the need across patient and population catchment areas (Washington, Coye, & Boulware, 2016).

This chapter describes the processes and lessons learned of the Duke Cancer Institute's (DCI) long standing community partnerships that led to Project PLACE (Population Level Approaches to Cancer Elimination), an NCI-funded community health assessment designed to identify, describe, and develop a community partnered research agenda to advance cancer equity within the DCI catchment area. The capacity to address disparities vary from one cancer center to another and the strategies outlined in this chapter can serve as a scalable example of a community outreach and engagement model to improve population and community health, particularly for the underserved. We illustrate the application of a community partnered health assessment model and offer examples of strategic collaborations, successes, challenges, lessons learned and implications for practice.

1.1 Burden of cancer and lack of diversity in clinical trials in North Carolina

There are 10,042,802 people living in NC, where African Americans make up the second largest race group at 22%. According to census data, the Latino/Hispanic population makes up 9% of the state population and has increased dramatically by over 300% in the last 10 years (US Census Bureau, 2018). Latinos/Hispanics have high concentrations in urban counties, such as Wake and in several rural counties where farming is one of the primary industries. Another growing and diverse community is the Asian population, although relatively small at 2.7%, this population has been steadily increasing over recent years. Native Americans are the smallest race/ethnic population in NC and are heavily concentrated in select communities such as Robertson County where the population is 39.7% Native American, 24.7% African American and 32.2% white. The average income in NC is $25,608, and 17.2% live at or below the federal poverty level. In terms of educational attainment, 85.4% have graduated high school, and 27.8% hold a bachelor's degree or higher (US Census Bureau, 2018). Across the state, there are a significant number of counties with a higher number of underrepresented race and ethnic minority groups compared to their majority white populations. For example, in Durham County, 51% of the population is comprised of underrepresented minorities, with 39% African American. Statewide studies show urban and rural differences in the cancer burden across the state where rural communities tend to fare worse overall. Similar to national population trends, race and ethnic differences exist where minorities in general have a lower incidence in cancer, yet are more likely to be diagnosed at later stages, and present with more aggressive disease. These trends differ in some cases where the burden can be even greater in the context of incidence and mortality. For example, African American men have a 1.6-fold higher incidence of prostate cancer, and a 2.4-fold higher mortality rate compared to their white counterparts (SEER, 2019).

Lack of minority participation in oncology clinical trials and biorepositories is a national problem. Current studies show that among the 15.5 million cancer survivors in the United States, only 9% are involved in clinical trials and even less are race and ethnic minorities, where current estimates range between 2% and 3% . Racial and ethnic minorities, the poor, and those who live in rural or low resourced communities are less likely to be involved in clinical research due to numerous factors including distance, fear, costs, or simply not being asked to participate (Baquet, Commiskey, Daniel Mullins, & Mishra, 2006; Dang et al., 2014; Paskett et al., 2002). In NC, a recent study found that men and minorities are least likely to participate in oncology clinical trials, highlighting the need to better understand these trends and develop mechanisms to address them (Hagiwara et al., 2014; Zullig et al., 2016). The under-representation of minorities and men in clinical research and trials has significant scientific implications. For example, from a biological standpoint, current studies are finding that black men are responding better to prostate cancer clinical trials than white men, underscoring the importance of diverse participation in clinical trials

(Amstrong, 2017; George et al., 2019; Halabi et al., 2018). Poor minority recruitment and retention efforts can compromise generalizability of research findings, raise concerns around biased reporting of adverse effects that may differ by race, and limit minorities and men from fully benefitting from research including access to cutting-edge and potentially life-enhancing clinical therapies (Unger, Cook, Tai, & Bleyer, 2016).

A myriad of individual, community, biological and system level barriers contribute to disparities in cancer, cancer care and research participation (Amstrong, 2017; Bandini, Preisser, & Nazzani, 2018; Barrett, Ingraham, Vann Hawkins, & Moorman, 2017; Durant et al., 2014; Ford et al., 2008). Lack of access to information, care, research, and resources, culture and communication including language, being under or uninsured, systemic racism and implicit bias in the health system, availability of cancer services and research, transportation, distrust of the medical establishment and research due to systemic racism and historical and broader social inequities all negatively impact cancer outcomes across the care continuum (Barrett et al., 2016; Durant et al., 2014; Hamel et al., 2016; Langford et al., 2014). Given higher rates of cancer morbidity and mortality among underserved populations, the growing diversity of our patient and community populations, lack of diversity in clinical trials participation, and the need for strategic collaborations and partnerships to reduce cancer disparities are heightened (Wallerstein & Duran, 2006; Washington et al., 2016).

1.2 The role of cancer centers and community health assessments to identify and address the need

Among many accountability measures, cancer centers are expected to identify and assess the needs in their communities, increase clinical research participation, and ensure patients do not fall through the gaps in care (Paskett & Hiatt, 2018; Tai & Hiatt, 2017). However, barriers that prohibit access and utilization of cancer services and research vary by community, requiring cancer centers to develop and implement strategic plans to assess and respond effectively to disparities within their catchment area. One strategy to assess and describe the cancer burden and respond to the need, is by conducting community health assessments and employing key methodologies of community engagement and partnerships (Barrett et al., 2016; Wilkins Consuelo & Alberti Philip, 2019).

The opportunity to identify and develop partnered strategies to advance health equity research are critical and timely. In 2017 the NCI heightened focus on the importance of community engagement and outreach to address cancer disparities across NCI-designated Cancer Centers and their catchment areas. This priority has encouraged a renewed focus on stakeholder engagement and the importance of authentic and impactful collaborations in improving cancer outcomes through community and population health assessments, health equity research, and interventions across the cancer spectrum (Hiatt et al., 2018).

1.3 Addressing cancer disparities: Community engagement and community health assessments

Community engagement encompasses methodologies designed to address health disparities by forming authentic and equally valued partnerships and collaborations between diverse stakeholders from the community, health care system including patients, and the research enterprise (Barrett et al., 2016; Gwede et al., 2012; Michener et al., 2012; Wallerstein & Duran, 2006). Within this context, the goals typically circulate around improving access and the delivery of health care, enhancing resources and tools to promote healthy behaviors, and increasing inclusion of diverse populations in biomedical research. A robust community engagement program within cancer centers can bolster community-engaged research and has the potential to provide a dynamic platform to enhance the capacity of academic health centers including cancer centers to meet the needs of the communities and patients they serve.

Using community engagement approaches can address health disparities in ways that traditional efforts cannot and can particularly reach the most underserved in the community and within the healthcare system by valuing the perspectives of diverse stakeholders toward the elimination of health inequity (Gwede et al., 2012; Wilkins Consuelo & Alberti Philip, 2019). Uniting individuals with diverse perspectives and experiences allows for thoughtful discussion and opportunities for better understanding issues of interest (Hiatt et al., 2018). Community organizations and partners can use relationships with local residents and patients to provide insight and support when developing and implementing programs and engaging in research including clinical trials (Rodriguez, Torres, & Erwin, 2013; Wallerstein, Minkler, Carter-Edwards, Avila, & Sánchez, 2015). Moreover, with authentic collaboration from the onset, research findings and programs intended to increase screening, ensure access and utilization of services by traditionally underrepresented groups, and diversifying clinical trial participation are more likely to be implemented (Barrett et al., 2017; O'Brien & Whitaker, 2011). Critical to the community engagement process is the importance of understanding diverse community perspectives and priorities around health. Essentially, understanding the nuances of community needs provides context for researchers as they engage in projects and disseminate findings (O'Brien & Whitaker, 2011; Wallerstein & Duran, 2006).

Several studies have developed community and academic partnerships to identify and address health disparities and have led to improved outcomes in African American and Latino/Hispanic populations, and in rural and urban communities (Barrett et al., 2017; García-Rivera et al., 2017; Meade, Menard, Luque, Martinez-Tyson, & Gwede, 2011). Such studies have shown improved or promising outcomes across a variety of chronic disease interventions including cancer, diabetes, HIV and cardiovascular conditions and highlight the importance of community engagement and partnerships to reduce health disparities. By employing community engagement methods to develop institution level strategic priorities around cancer health equity research, cancer centers have the opportunity to reach the most underserved in the community and within the healthcare system in ways typically not fully afforded by traditional strategic activities. As such, community engagement is paramount to the development and implementation of community health needs assessments, and the subsequent research priorities and interventions necessary to address cancer health disparities within cancer center catchment areas (Barrett et al., 2017).

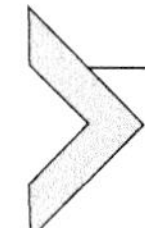

2. Methods

2.1 Building infrastructure and a health equity agenda through community engagement and partnerships

In 2012, the DCI, established the Office of Health Equity (OHE) to develop a health disparities and equity agenda to reduce cancer disparities. Two key factors triggered the prioritization of health disparities as a key strategy within the DCI. First, under the new leadership the DCI embarked on a new model of research and patient care through a coordinated effort to authentically engage the community in outreach, screening, and research as both participants and experts to inform the research process (Barrett et al., 2017).

The second factor that triggered prioritization was that accrediting entities and funding agencies were implementing policies which heighten accountability around community engagement and health disparities and equity efforts for cancer centers (Barrett et al., 2017; Hiatt et al., 2018). As a result, three areas of focus were established and include:

- Greater emphasis on the use of community assessments to understand and effectively respond to community and patient needs.
- Increased accountability in minority accrual and retention in research and clinical trials.
- The importance of engaging patients and community partners in health disparities research.

Through collaborative community partnerships the DCI, OHE convened a Community Advisory Council and conducted a qualitative cancer health assessment reaching over 230 participants. The assessment led to a co-created and executed community and academic health disparities strategic plan which highlights a health system/academic and community partnered platform to serve the community, patients, researchers, and clinicians.

DCI collaborated with community based organizations and other key stakeholders to develop a dynamic integrated infrastructure complete with initiatives to improve cancer prevention and control, increase early detection, and address cancer disparities and gaps in access to care and research within the region. This process resulted in the development of a Community Facing Navigation Program with permanent health system supported positions (not dependent on external grants), and the creation of several key outreach programs which include free community-based screening programs, community outreach and education on cancer, and biomedical research participation (Barrett et al., 2017). Building on the established infrastructure, in 2017 the DCI worked with its Community Advisory Council (CAC), as well as with additional well-established community partners, to develop and implement an NCI-funded community health assessment, entitled Project PLACE (see Fig. 1). Project PLACE is a quantitative community health assessment designed to inform and shape the roadmap for strategic research, outreach and interventions to reduce cancer disparities, and to engage in targeted efforts to increase and diversify clinical research participation within the catchment area.

2.2 Development and implementation of Project PLACE: An academic and community partnered health assessment

Funded by an NCI P30 Supplement to define the cancer burden of its catchment area, the overarching goal of Project PLACE was to extend DCI's current infrastructure beyond community engagement and collaborations to include a robust data platform designed to inform and shape the roadmap for strategic research, outreach and interventions, and to engage in

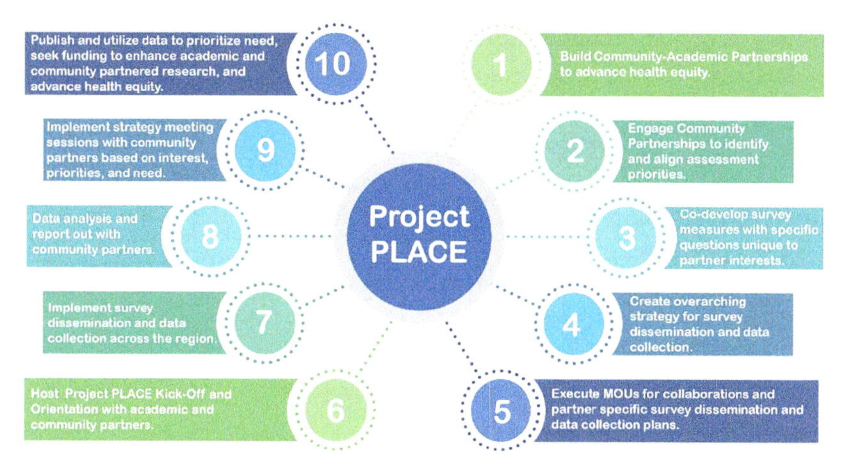

Fig. 1 Partnership development, survey dissemination, data collection, and reporting process.

targeted efforts to increase clinical research participation. Salient to this venture is both understanding and describing key factors impacting cancer disparities across diverse populations, aligning the priorities of research teams and partnering organizations to build capacity through strategic collaborations to address cancer disparities, spur innovative research questions and the development of comprehensive and multi-pronged strategies to increase and diversify clinical trial participation.

2.3 Engaging the DCI Community Advisory Council to guide and inform Project PLACE

The DCI's CAC is a dynamic and vital component of the health disparities work within the DCI. The council is comprised of 22 individuals representing diverse perspectives across the cancer spectrum. Collectively, the council is comprised of educators, health professionals, researchers, faith leaders, grass-roots organizers, cancer survivors/patients, community advocates, and more, while representing diversity across race, ethnicity, class, religion, geography, sexuality/identity, and many other perspectives. These partners access and engage their broader community constituency based on identified programming and research priorities. The committee serves in an advisory and collaborative capacity to provide guidance, feedback, development and execution of research projects at the DCI, and promote community and outreach programming to increase understanding and involvement in clinical trials and research programs among underrepresented groups, and improve access and the delivery of cancer screening and care across the continuum of care. The CAC plays an instrumental role in the development, execution, and outcomes of Project PLACE. In addition, the DCI has a broader network of community partners that represent national, statewide regional, and local health organizations, community-based and rooted organizations, local historically black colleges and universities (HBCUs), along with cancer patients and caregivers.

2.4 Survey development: Aligning NCI, DCI, and community priorities

The survey development process of Project PLACE had to incorporate priorities and guidelines from national, regional, and local stakeholders interested in the outcomes and subsequent research, community engagement, and interventions. The DCI was one of 15 grantees who received NCI funding to describe their respective catchment areas with the plan of using the data to ultimately share strategic activities including community outreach, increase diversity in clinical trials and address patient care with a cross cutting theme of addressing disparities. NCI, in collaboration with other stakeholders including the P30 supplement grantees, selected five core measures from the Health Information National Trends Survey (HINTS) data set and all sites were expected to capture these data through their local community assessments (Gage-Bouchard & Rawl, 2019). The DCI CAC, partners in the network, and researchers across a variety of cancer related disciplines identified an additional eight DCI-specific measures to incorporate into the survey. The final survey included 91 items with a total of 13 measures and was available in English, Spanish and Chinese (see Table 1).

In addition, each community organization was asked if there were questions that were important to them for building capacity in their own organization that extends beyond the agreed upon measures already selected for the survey. Two partnering organizations chose to add additional questions to the instrument that were pertinent to their specific stakeholder populations and important for their own capacity building. Those items were added as an addendum to the survey. The survey was constructed using the DCI's Behavioral Science and Survey Research Shared Resource, supported by the NCI Cancer Center Support Grant (P30CAO14236 45).

2.4.1 Identifying community partners and collaborators

DCI engaged long standing and newly established organizations and partners to develop a survey dissemination plan that would reach diverse populations cutting across key demographics including race, ethnicity, age, region,

Table 1 Project PLACE core measures.

NCI core measures captured across all NCI-funded sites
1. Access to care
2. Demographics
3. Accessing health information
4. Tobacco use
5. Cancer screening and knowledge
DCI and community specific core measures
6. Medical research participation
7. Genetic testing
8. Beliefs about cancer
9. Medical mistrust
10. Health history
11. Awareness and use of palliative care
12. Physical activity and exercise
13. Nutrition

culture, and socioeconomic status. We took a novel approach in this process by partnering and incentivizing organizations rather than the individual participants. This was critical as we aimed to not just partner for the purpose of capturing surveys but to review the outcome and ultimately identify opportunities for outreach, interventions, and community-engaged research. Partners and collaborators represented diverse communities, regions, and sociodemographic groups. The reach of Project PLACE partners includes a diverse population representing the African American/Black, Asian, Latinx/Hispanic, White, Muslim/Christian, Lesbian, Gay, Bisexual, Transsexual, Queer, Intersex, and Asexual communities and respondents from both rural/urban communities across the state. Fourteen community-based organizations represented a variety of groups including fraternities and sororities, health and patient advocacy groups, the LGBTQ centers, an HBCU, and senior centers. We partnered with five (5) faith organizations which included Catholic, non-Catholic Christian, and Muslim organizations. Last, a free health clinic serving uninsured patients spanning four counties in NC, and the Duke Health system provided the opportunity to engage a diverse patient population.

2.5 Community partner kick-off training

A Project PLACE kick-off meeting was held with community partners involved in the development and dissemination of the survey.

The 2-hour session provided an overview of the cancer burden and need in the community, the purpose of Project PLACE and the importance of partnerships and collaborations. Each organization shared details about their programs and services, who they serve and their reach. The kick-off then had an in-depth workshop that covered key survey operations such as the development and refinement of each partner's Survey Dissemination Plan, a review of research ethics, and Standard Operating Procedures (SOP's) to ensure activities were in full alignment with Duke's Institutional Review Board (IRB) and human subjects' protections.

2.6 Survey dissemination and compensation

The initial goal was to enroll 2000 participants across all racial and geographical groups into the study. The study was conducted within various com munities, with the majority of participants recruited via health fairs and outreach endeavors sponsored by community partners. The surveys were self-administered via either paper-pen format or online using a Research Electronic Data Capture (REDCap) database. The online survey link was distributed through selected community partners and to those who request to disseminate the online survey to their constituents. Most partners disseminated pen and paper surveys which were disseminated in a variety of settings including community forums, health fairs, and outreach events, during and after church services. We employed purposeful sampling through community-based convenience sampling methods by focusing on recruiting participants from community organizations based in Durham, Wake, Vance, Granville, Alamance, and Orange counties. We provided compensation ($10 per survey up to $2000) to partner organizations as part of our agreement to access the data and co-develop follow-up interventions and research in which the funds could be used to offset the costs of having staff distribute and collect surveys, provide participant incentives, or to meet a specific programmatic need in their organization. DCI OHE provided planning and coordination support as needed and OHE Community Navigators attended and helped facilitate the programs. In addition, participants that completed the survey at community partner events were offered gift items valued less than $5 such as a water bottle or a tote bag. DCI OHE provided planning and coordination support as needed and OHE Community Navigators attended the programs to answer questions.

The community health assessment posed no more than minimal risk to the participants and we were granted a waiver of consent from the DCI's IRB. In lieu of a consent process, a member of the research team provided a thorough review of the project details with instructions to complete the survey to the community partners as well as the participants completing the survey. The survey did not collect Personal Health Information (PHI) and could not be traced back to an individual. Research team members addressed any questions or concerns of the participants prior to them completing the survey.

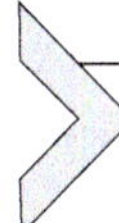

3. Project PLACE results and outcomes

3.1 Data collection outcomes by organization type

Project PLACE data collection spanned from April 2017 to December 2017. A total of 2315 surveys were completed, exceeding our goal of 2000. Most participants were recruited from community organizations primarily located in Durham, Wake, Vance, Alamance, and Johnston counties in Central NC, representing 24 diverse organizations including clinics, community-based organizations, faith organizations and community outreach events and programs (see Fig. 2).

Fourteen community organizations held 29 events capturing a total of 1119 surveys. We partnered with one health clinic that held an event capturing 152 surveys. Five faith organizations held 11 events and captured a total of 795 surveys. Programs at faith organizations varied from bible studies, health fairs, church services and Eid Al-Fitr within a Muslim mosque. We also partnered with organizations to host two annual health screenings and outreach programs targeting men and women in Durham and the

Fig. 2 Project PLACE: survey collection outcomes.

Greater Triangle region. Programs included the Women's Health Awareness Day, the Men's Health Initiative, Sister's Network Tea for Two, the LGBTQ Center's 2nd Anniversary Celebration, the NC Cancer Prevention & Control Branch Survivorship Summit, and the Lung Cancer Initiative Summit. Collectively the outreach programs collected 249 surveys. It is important to note, that 10 Project PLACE partners were a direct result of the collaborations with, or were facilitated by members of the DCI Community Advisory Council In summary, our 24 partners held 47 events reaching 2315 diverse respondents (see Table 2).

3.2 Survey outcomes and demographics

The Project PLACE partnership led to 2315 respondents completing surveys in diverse community settings. Surveys were completed in three languages: English (88%), Spanish (9.7%), and Mandarin (2.1%). Survey respondents represent significant racial and ethnic diversity. African Americans and whites had similar representation at 36% (840), and 37% (850), respectively. Asians made up 10% of the survey respondents and Native Americans the remaining 2%. Regarding ethnicity, 14% of the sample identified as Hispanic. Females were 61% of the respondents. Six percent of the respondents identified as a sexual minority (lesbian, gay, and bisexual) and 84% identified as straight/heterosexual. Twenty-eight of the sample had a vocational training or less, and 46% had a college degree or higher. The average age of participants was 51 with a range from 18 to 99 years old. Eighty-one percent had health insurance (public or private), and 78% were in metro/urban communities whereas 17% were from non-metro/rural communities (see Table 3).

Table 2 Survey collection by type of organization.

Organization type	Number of organizations	Total surveys collected	Total # events
Community-based	14	1119	29
Faith organization	5	795	11
Health clinic	1	152	1
Community outreach	4	249	6
Total	24	2315	47

Table 3 Project PLACE demographics ($N=2315$).

Demographic characteristics		**N**	**% or mean**
Race	White	850	37
	African American/Black	840	36
	Asian/Pacific Islander	243	10
	Native American	37	2
	More than 1 race	89	12
Ethnicity	Hispanic	326	14
	Non-Hispanic	1739	75
Gender	Male	783	34
	Female	1402	61
	Other	8	0
Sexual orientation/ gender identity	Straight/heterosexual	1941	84
	Homosexual/gay	81	4
	Transgender, transsexual, or gender non-conforming	10	0
	More than 1 selected	12	1
Education	High-school or less	519	22
	Some college or technical school	606	26
	College graduate	466	20
	Post-graduate	148	6
Geographic location	Metro	1802	78
	Non-metro	401	17
Mean age			51
Health insurance status[a]	Yes	1875	81
	No	281	12
Survey language	English	2042	88
	Spanish	225	9.7
	Mandarin	8	2.1

[a]Do you have any kind of healthcare coverage, including health insurance, prepaid plans such as HMOs, or government plans such as Medicare, or Indian Health Service?

3.3 Project PLACE outcomes: Stakeholder report outs and partnered strategies

An important aspect of community partnered and engaged research is to ensure findings and outcomes are reported back to the community in a meaningful and timely manner. Consistent with this important principle of engagement, the DCI co-presented the data and outcomes for the individual organizations and the whole sample. A total of 12 report back sessions were held, reaching over 1200 people in the community. Where some partners wanted to meet with a small leadership team to discuss highlights of the findings, others requested a full co-presentation. All partners wanted to be connected to resources to promote access to care and awareness through the Community Facing Navigation Program and about 50% ($n = 12$) were ready to conduct some level of intervention or program based on the findings from their constituents. To date, eight Health Equity Strategy Groups have formed around Project PLACE data and are at various stages of productivity ranging from publications, small grants, further community data collection through focus groups and listening sessions, and planning fundable interventions using the Project PLACE data for preliminary data. Strategy group topics include: Hospice and Palliative Care, Women's Cancer and Screening Behaviors, Prostate, and Colorectal Cancer Screening Knowledge and Behaviors, Genetic Testing and Cancer Diagnosis and History of Cancer, Rural Health and Cancer, Race Disparities in Biomedical Research Participation, Medical Mistrust and Research Participation, Community-Based Screening Program and Biomedical Research Participation, Health Information, Communication, and Technology, and Promoting Men's Health and Reducing Cancer Risk.

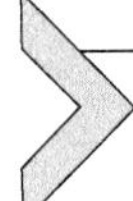

4. Discussion

4.1 Project PLACE successes

Project PLACE has several key strengths that should be noted. Specifically, the success of the community health assessment process is noted in several distinct but interrelated ways. First, building on DCI's extensive and long standing relationships and partners, over 2300 self-administered pen and paper surveys from a very diverse population were captured within a 7-month period. This highlights the impact of long standing and mutually beneficial community engagement when academic health centers and cancer centers are seeking to better understand, partner, and serve their constituents within the catchment area. Second, the survey was a 91-item tool and yet did not pose a collective barrier to completing the survey across a variety of settings. Providing the survey in three languages may have added to making this survey completion more reasonable for respondents.

Trust is another key aspect of this study. In most cases, community members were already familiar with the DCI from outreach activities related to access to care, cancer screening and clinical trials participation. In a few instances the DCI had minimal engagement with the organization, but worked through a "trust broker," a community leader or partner who served as the liaison to the community. This led to a significantly higher rate of participation and solidified the opportunity to conduct the subsequent report out sessions and next steps in regards to services, research, and interventions. Another key aspect that cannot be underestimated is the well-established DCI CAC. Almost 50% of the community organizations represented on the CAC were part of the Project PLACE assessment, and collectively the group provided guidance from survey development through, implementation, and report out to the community, and the next step strategies. The richness of this engagement influences both the direct work that stems from Project PLACE, and the ongoing and broader activities of the DCI.

Lastly, Project PLACE not only allowed the DCI to leverage existing relationships, it also opened the door for new ones. These new relationships create opportunities to expand the diversity of community partners and constituents. For example, through this assessment process DCI began new relations with a local mosque, a Chinese community-based organization, and some rural based senior centers. New partners were a part of the kick-off training and celebration and allowed them to see their participation as being part of a broader community, and highlighted the successes, resources, and programs that came from the first strategic assessment conducted 5 years prior, and the impact it has made on reaching diverse and underserved populations in the community.

Two additional important aspects of this program are (1) the model for incentivizing the community organization instead of the individuals and (2) the opportunity for community partnering organizations to add their own questions to the survey that can help build their capacity to identify and meet the needs of their constituents. Community organizations designed and led the survey dissemination aspects of the program with the plan to use the data to co-publish and co-develop research and programs to improve cancer outcomes with the DCI. Consequently, community organizations were compensated for the surveys as part of the development of a strategic goal and plan using the data once the survey was complete. Likewise, as partners we capitalize on the survey process by ensuring space for community organizations to capture data that was specifically important to their mission and capacity to serve their constituents. This was important as we work to create open, transparent, and mutually beneficial opportunities within academic and community partnerships.

4.2 Lessons learned

The success of the program does not come without key lessons learned. The online survey link yielded very few respondents and overall was not a successful mechanism to engage community partners and their constituents. Although survey collection methods often use email to reach populations in research, given our target audience, this did not have the intended impact our partners or the DCI anticipated. It seems the pen and pencil administration of the survey had an added effect of engaging people and meeting them where they were to conduct the assessment. Face-to-face administration is also critical for relationship building and follow-up activities that will result from the survey findings. These two points should be underscored as they demonstrate key aspects to meaningful and authentic community engagement.

5. Conclusion

Project PLACE illustrates the impact stakeholder engagement and robust community partnerships can have when conducting community and population health assessments. Implementing community health assessments where partners share the development, dissemination, and subsequent research and programs based on the findings are important and critical to success and sustainability of programs, research, and interventions. Project PLACE created key opportunities to consider when partnering to understand and address the needs in a given community. Ensuring there is value and mutually beneficial outcomes by leveraging activities to meet the capacity building needs of the community partners as they work with the cancer center, and on their own to address key aspects of their mission, is important and should not be overlooked. Moreover, using assessment data to collaboratively co-develop programs, research, and services to improve population health and advance cancer health equity is an excellent model toward building trust, and community capacity to address cancer disparities, specifically within comprehensive cancer center catchment areas.

Acknowledgments

This research was supported by NCI/NIH Grant Number P30 CA014236. Special appreciation to the following people who served on either the Research Council, the Community Advisory Council for the Duke Cancer Institute (DCI), and/or as collaborators or supporters of Project PLACE (Population Level Approaches to Cancer Elimination): Sue McLaurin, Ava Crawford (NCDHHS), Debi Nelson (NCDHHS), Women's Health Awareness Day (an outreach program developed and supported by National Institute of Environmental Health Sciences), Pao Hwa Lin, Pilar Rocha Rosenberg, Maritza Chirinos, Helena Cragg, Patricia Wigfall, Ping Zhang, Jenny Denai, Eric Ireland, Bo Marshall, Michael Palmer, Claudia Graham, Pastor Raj, Kerri Burnette, Awanya Davis, Marsha Edwards, Reverend Jerome Taylor, Ron Sangal, Kenisha Bethea, Steven Patierno, Terry Hyslop, Xiaomei Gao, Patricia Moorman, Charmaine Royal, Qingyi Wei, Kevin Oeffinger, Devon Noonan, Kathryn Pollack, Demetrius Harvey, and Jeffrey Ford.

References

Amstrong, A. J. (2017). Prognostic and predictive biomarkers in metastatic castration-resistant prostate cancer. *Clinical Advances in Hematology & Oncology, 15*(3).

Baquet, C. R., Commiskey, P., Daniel Mullins, C., & Mishra, S. I. (2006). Recruitment and participation in clinical trials: Socio-demographic, rural/urban, and health care access predictors. *Cancer Detection and Prevention, 30*(1), 24–33.

Bandini, M., Preisser, F., Nazzani, S., et al. (2018). The effect of other-cause mortality adjustment on access to alternative treatment modalities for localized prostate cancer among African American patients. *European Urology Oncology, 1*(3), 215–222. https://doi.org/10.1016/j.euo.2018.03.007.

Barrett, N. J., Hawkins, T. V., Wilder, J., Ingraham, K. L., Worthy, V., Boyce, X., et al. (2016). Implementation of a health disparities & equity program at the Duke Cancer Institute. *Oncology Issues, 31*(5), 48–57.

Barrett, N. J., Ingraham, K. L., Vann Hawkins, T., & Moorman, P. G. (2017). Engaging African Americans in research: The recruiter's perspective. *Ethnicity & Disease, 27*(4), 453–462.

Dang, J. H., Rodriguez, E. M., Luque, J. S., Erwin, D. O., Meade, C. D., & Chen, M. S., Jr. (2014). Engaging diverse populations about biospecimen donation for cancer research. *Journal of Community Genetics, 5*(4), 313–327.

Durant, R. W., Wenzel, J. A., Scarinci, I. C., Paterniti, D. A., Fouad, M. N., Hurd, T. C., et al. (2014). Perspectives on barriers and facilitators to minority recruitment for clinical trials among cancer center leaders, investigators, research staff, and referring clinicians: Enhancing minority participation in clinical trials (EMPaCT). *Cancer, 120*(Suppl. 7), 1097–1105.

Ford, J. G., Howerton, M. W., Lai, G. Y., Gary, T. L., et al. (2008). Barriers to recruiting underrepresented populations to cancer clinical trials: A systematic review. *Cancer, 112*(2), 228–242.

Gage-Bouchard, E. A., & Rawl, S. (2019). Standardizing measurement of social and behavioral dimensions of cancer prevention and control to enhance outreach and engagement in NCI-designated cancer centers. *Cancer Epidemiology, Biomarkers and Prevention, 28*, 431–434. https://doi.org/10.1158/1055-9965.EPI-18-0794.

García-Rivera, E. J., Pacheco, P., Colón, M., Mays, M. H., Rivera, M., Munet-Díaz, et al. (2017). Building bridges to address health disparities in Puerto Rico: The "Salud para Piñones" project. *Puerto Rico Health Sciences Journal, 36*(2), 92–100.

George, D., Heath, E. I., Sartor, A. O., Sonpavde, G., Berry, W. R., Healy, P., et al. (2019). Abi race: A prospective, multicenter study of black (B) and white (W) patients (pts) with metastatic castrate resistant prostate cancer (mCRPC) treated with abiraterone acetate and prednisone (AAP). *Journal of Clinical Oncology*, *36*(18). https://doi.org/10.1200/JCO.2018.36.18_suppl.LBA5009.

Gwede, C. K., Castro, E., Brandon, T. H., McIntyre, J., Meade, C. D., Munoz-Antonia, T., et al. (2012). Developing strategies for reducing cancer disparities via cross-institutional collaboration: Outreach efforts for the partnership between the Ponce School of Medicine and the Moffitt Cancer Center. *Health Promotion Practice*, *13*(6), 807–815.

Hagiwara, N., Berry-Bobovski, L., Francis, C., Ramsey, L., Chapman, R. A., & Albrecht, T. L. (2014). Unexpected findings in the exploration of African American underrepresentation in biospecimen collection and biobanks. *Journal of Cancer Education: The Official Journal of the American Association for Cancer Education*, *29*(3), 580–587.

Halabi, S., Dutta, S., Tangen, C. M., Rosenthal, M., Petrylak, D. P., Thompson, I. M., et al. (2018). Overall survival between African-American (AA) and Caucasian (C) men with metastatic castration-resistant prostate cancer (mCRPC). *Journal of Clinical Oncology*, *36*(18 Suppl), LBA5005. https://www.ncbi.nlm.nih.gov/pubmed/30576268. PMID: 30576268.

Hamel, L. M., Penner, L. A., Albrecht, T. L., Heath, E., Gwede, C. K., & Eggly, S. (2016). Barriers to clinical trial enrollment in racial and ethnic minority patients with cancer. *Cancer Control*, *23*(4), 327–337.

Hiatt, R. A., Sibley, A., Fejerman, L., Glantz, S., Nguyen, T., Pasick, R., et al. (2018). The San Francisco cancer initiative: A community effort to reduce the population burden of cancer. *Health affairs (Project Hope)*, *37*(1), 54–61.

Langford, A. T., Resnicow, K., Dimond, E. P., Denicoff, A. M., Germain, D. S., McCaskill-Stevens, W., et al. (2014). Racial/ethnic differences in clinical trial enrollment, refusal rates, ineligibility, and reasons for decline among patients at sites in the National Cancer Institute's Community Cancer Centers Program. *Cancer*, *120*(6), 877–884.

Meade, C. D., Menard, J. M., Luque, J. S., Martinez-Tyson, D., & Gwede, C. K. (2011). Creating community-academic partnerships for cancer disparities research and health promotion. *Health Promotion Practice*, *12*(3), 456–462.

Michener, M., Cook, J., Ahmed, S. M., Yonas, M. A., et al. (2012). Aligning the goals of community-engaged research: Why and how academic health centers can successfully engage with communities to improve health. *Academic Medicine*, *87*(3), 285–291.

O'Brien, M. J., & Whitaker, R. C. (2011). The role of community-based participatory research to inform local health policy: A case study. *Journal of General Internal Medicine*, *26*, 1498–1501. https://doi.org/10.1007/s11606-011-1878-3.

Paskett, E. D., Cooper, M. R., Stark, N., Ricketts, T. C., Tropman, S., Hatzell, T., et al. (2002). Clinical trial enrollment of rural patients with cancer. *Cancer Practice*, *10*(1), 28–35.

Paskett, E. D., & Hiatt, R. A. (2018). Catchment areas and community outreach and engagement: The new mandate for NCI-designated Cancer Centers. *Cancer Epidemiology, Biomarkers & Prevention: A Publication of the American Association for Cancer Research, Cosponsored by the American Society of Preventive Oncology*, *27*(5), 517.

Rodriguez, E. M., Torres, E. T., & Erwin, D. O. (2013). Awareness and interest in biospecimen donation for cancer research: Views from gatekeepers and prospective participants in the Latino community. *Journal of Community Genetics*, *4*(4), 461–468.

SEER. (2019). *SEER facts sheets: Disparities*. https://seer.cancer.gov/statfacts/html/disparities.html. Available online at: Last accessed May 5, 2019.

Tai, C. G., & Hiatt, R. A. (2017). The population burden of cancer: Research driven by the catchment area of a cancer center. *Epidemiologic Reviews*, *39*(1), 108–122.

Unger, J. M., Cook, E., Tai, E., & Bleyer, A. (2016). The role of clinical trial participation in cancer research: Barriers, evidence, and strategies. *American Society of Clinical Oncology Educational Book. American Society of Clinical Oncology Meeting, 35*, 185–198.

US Census Bureau. (2018). *Washington DC: US Census Bureau*. [Available from: https://www.census.gov/en.html Accessed May 5, 2019].

Wallerstein, N., & Duran, B. (2006). Using community based participatory research to address health disparities. *Health Promotion Practice*, 7(3), 312–323.

Washington, A. E., Coye, M. J., & Boulware, L. E. (2016). Academic health systems' third curve population health improvement. *JAMA, 315*(5), 459–460.

Wilkins Consuelo, H., & Alberti Philip, M. (2019). Shifting Academic Health Centers from a culture of community service to community engagement and integration. Engagement and integration. Published online ahead of print March 19 *Academic Medicine, 94*, 763–767. https://doi.org/10.1097/ACM.0000000000002711.

Further reading

Barrett, N. J., Rodriguez, E. M., Iachan, R., et al. (2020). Factors associated with biomedical research participation within community-based samples across 3 National Cancer Institute-designated cancer centers. *Cancer, 126*(5), 1077–1089. https://doi.org/10.1002/cncr.32487.

Dang, J. H. T., & Chen, M. S., Jr. (2018). Time, trust, and transparency: Lessons learned from collecting blood biospecimens for cancer research from the Asian American community. *Cancer, 124*(Suppl. 7), 1614–1621.

Fouad, M. N., Johnson, R. E., Nagy, M. C., Person, S. D., & Partridge, E. E. (2014). Adherence and retention in clinical trials: A community-based approach. *Cancer, 120*(Suppl. 7), 1106–1112.

Haring, R. C., Henry, W. A., Hudson, M., Rodriguez, E. M., & Taualii, M. (2018). Views on clinical trial recruitment, biospecimen collection, and cancer research: Population science from landscapes of the Haudenosaunee (People of the Longhouse). *Journal of Cancer Education, 33*(1), 44–51.

Langford, A., Resnicow, K., & An, L. (2010). Clinical trial awareness among racial/ethnic minorities in HINTS 2007: Sociodemographic, attitudinal, and knowledge correlates. *Journal of Health Communication, 15*(Suppl. 3), 92–101.

Langford, A. T., Resnicow, K., & Beasley, D. D. (2015). Outcomes from the body & soul clinical trials project: A university-church partnership to improve African American enrollment in a clinical trial registry. *Patient Education and Counseling, 98*(2), 245–250.

Llanos, A. A., Young, G. S., Baltic, R., Lengerich, E. J., Aumiller, B. B., Dignan, M. B., et al. (2018). Predictors of willingness to participate in biospecimen donation and biobanking among Appalachian adults. *Journal of Health Care for the Poor and Underserved, 29*(2), 743–766.

Murthy, V. H., Krumholz, H. M., & Gross, C. P. (2004). Participation in cancer clinical trials: Race-, sex-, and age-based disparities. *JAMA, 291*(22), 2720–2726.

N.C. Division of Public Health. (2019). Cancer Control Branch, Reducing the cancer burden in North Carolina. https://publichealth.nc.gov/chronicdiseaseandinjury/cancerpreventionandcontrol/docs/ReducingtheBurdenofCancerResourceGuide.pdf. Available online at: Retrieved: April 5, 2019.

NC State Cancer Profile. (2019). North Carolina State cancer profile. https://statecancerprofiles.cancer.gov/quick-profiles/index.php?statename=northcarolina. Available online at: Last Accessed May 2, 2019.

Newman, L. A., Roff, N. K., & Weinberg, A. D. (2008). Cancer clinical trials accrual: Missed opportunities to address disparities and missed opportunities to improve outcomes for all. *Annals of Surgical Oncology, 15*(7), 1818–1819.

National Cancer Institute: Surveillance, Epidemiology and End Results Program, Prostate Cancer Stat Fact, 2016. http://seer.cancer:gov/statfacts/html/prost.html, 2016: Last accessed July 13, 2016.

Rodriguez, E. M., Saad-Harfouche, F. G., Miller, A., Mahoney, M. C., Ambrosone, C. B., Morrison, C. D., et al. (2016). Engaging diverse populations in biospecimen donation: Results from the Hoy y Manana study. *Journal of Community Genetics*, 7(4), 271–277.

Stewart, J. H., Bertoni, A. G., Staten, J. L., Levine, E. A., & Gross, C. P. (2007). Participation in surgical oncology clinical trials: Gender-, race/ethnicity-, and age-based disparities. *Annals of Surgical Oncology*, *14*(12), 3328–3334.

Unger, J. M., Hershman, D. L., Albain, K. S., Moinpour, C. M., Petersen, J. A., Burg, K., et al. (2013). Patient income level and cancer clinical trial participation. *Journal of Clinical Oncology: Official Journal of the American Society of Clinical Oncology*, *31*(5), 536–542.

Vanderpool, R. C., Kornfeld, J., Mills, L., & Byrne, M. M. (2011). Rural-urban differences in discussions of cancer treatment clinical trials. *Patient Education and Counseling*, *85*(2), e69–e74.

Wallerstein, N., Minkler, M., Carter-Edwards, L., Avila, M., & Sánchez, V. (2015). Improving health through community engagement, community organization, and community building. In K. Glanz, B. K. Rimer, & K. Viswanath (Eds.), *Health behavior: Theory, research, and practice* (pp. 277–300). Jossey-Bass.

Zullig, L. L., Fortune-Britt, A. G., Rao, S., Tyree, S. D., Godley, P. A., & Carpenter, W. R. (2016). Enrollment and racial disparities in cancer treatment clinical trials in North Carolina. *North Carolina Medical Journal*, 77(1), 52–58. https://doi.org/10.18043/ncm.77.1.52.

CHAPTER NINE

Mighty men: A faith-based weight loss intervention to reduce cancer risk in African American men

Derek M. Griffith[a,b,*], Emily C. Jaeger[a]
[a]Center for Research on Men's Health, Vanderbilt University, Nashville, TN, United States
[b]Center for Medicine, Health and Society, Vanderbilt University, Nashville, TN, United States
[*]Corresponding author: e-mail address: derek.griffith@vanderbilt.edu

Contents

Abstract

According to the American Cancer Society's guidelines on nutrition and physical activity for cancer prevention, weight control, eating practices and physical activity are second only to tobacco use as modifiable determinants of cancer risk. However, no evidence-based interventions have been targeted to African American men or tailored to individual African American men's preferences, needs or identities. The goal of this chapter is to describe the rationale for the components, aims and setting of Mighty Men: A Faith-Based Weight Loss Intervention for African American Men. We begin by discussing the rationale for focusing on weight loss in the context of cancer prevention, and argue that obesity and obesogenic behaviors are important yet modifiable determinants of cancer risk. Next, we briefly review the scarce literature on interventions to promote healthy eating, physical activity and weight loss in our population of interest, and then discuss the rationale for conducting the intervention in faith- based organizations rather

Advances in Cancer Research, Volume 146
ISSN 0065-230X
https://doi.org/10.1016/bs.acr.2020.01.010

than other common settings for recruiting African American men. We conclude with a discussion of the conceptual foundations and components of Mighty Men, and discuss our focus and goals in the context of the larger literature in this area.

1. Introduction

Men, especially African American men, regularly engage in over 30 different behaviors that increase their risk for morbidity and mortality, including behaviors that are often implicated in explanations of differences in cancer risk between men and women (e.g., unhealthy eating, sedentary behavior, avoiding cancer screening, tobacco use, HPV transmission) (Courtenay, 2000, 2002; Griffith, Gunter, & Allen, 2011; Peterson, 2009; Robertson, 2007). These behaviors often are also culturally sanctioned ways of distinguishing among males and between males and females, but are rarely considered in interventions for African American men (Courtenay, 2000; Evans, Frank, Oliffe, & Gregory, 2011). Since publication of the 10-volume Secretary's Task Force on Black and Minority Health in 1985, it has been clear that African American men experience poor health outcomes, particularly in terms of cancer mortality (Jack & Griffith, 2013).

Part of the challenge in addressing cancer disparities among men has been that African American men's health has typically been examined by focusing on racial and ethnic health disparities, with little attention to gender (Jack & Griffith, 2013). While useful for identifying environmental and cultural factors associated with race and ethnicity, this strategy has failed to achieve an equal number of African American men participants in part because the interventions do not address gender in their design and materials nor do they attend to the role that gender plays in health behavior or men's lives (Griffith, Gunter, & Allen, 2012).

In this chapter, we describe the rationale, context and justification for *Mighty Men: A Faith- Based Weight Loss Intervention for African American Men*. The focus and goals of this intervention are consistent with the National Institute of Minority Health and Health Disparities definition of minority health research, "...the scientific investigation of these distinctive health characteristics and attributes of minority racial and/or ethnic groups who are usually underrepresented in biomedical research in order to understand population health outcomes" (NIMHD, 2017). We highlight how *Mighty Men* is an effort to promote the health of African American men.

Additionally, we focus on *Mighty Men* as a representative model for how to build on the strengths of the African American community (e.g., faith-based organizations) and integrate positive notions of how the men involved view what it means to be an African American man (Griffith, Allen, et al., 2010; Griffith, Campbell, Allen, Robinson, & Stewart, 2010; Griffith & Cornish, 2018).

2. Background

2.1 Cancer risk in African American men

Men in the United States (U.S.) have a one in two lifetime risk of developing cancer, and that risk is significantly higher for African American men (Griffith & Johnson, 2012; Griffith et al., 2007; American Cancer Society, 2013a, 2013b). For African American men aged 45 and older, cancer is one of the top two leading causes of death (American Cancer Society, 2013a, 2013b; Taylor et al., 2001), with African American men having the highest rates of cancer mortality among racial/ethnic groups of men or women in the U.S. (American Cancer Society, 2009, 2013a, 2013b; Griffith & Johnson, 2012). Obesity is associated with prostate, lung, and colon cancers, and these cancers comprise 60% of all cancer incidence and 54% of all cancer mortality for African American males (American Cancer Society, 2009). African American men's higher rates of cancer risk, morbidity, mortality and obesity seem to be associated with their lower rates of protective health behaviors (American Cancer Society, 2013a, 2013b; Griffith & Johnson, 2012).

2.2 Obesity: A Key, modifiable determinant of cancer risk for African American men

Obesity—defined as a body mass index (BMI) of 30.0 kg/m^2 or higher (Centers for Disease Control and Prevention, 2016)—is an important, modifiable determinant of cancer disparities and cancer risk. Obesity is one of the leading yet most preventable causes of colorectal cancer, lung cancer, aggressive prostate cancer (American Cancer Society, 2013a, 2013b; Hebert et al., 2012; Key et al., 2004; Koh, Massin-Short, & Elqura, 2009; Najari et al., 2012; Wei et al., 2004) and overall cancer disparities. The key mechanism that contributes to obesity is an unhealthy "energy balance," or when the calories consumed in eating and drinking exceed the calories expended in physical activity (Hill, 2006). Obese people who intentionally lose weight through increased healthy eating and physical activity showed reduced rates of hormones related to cancer risk including insulin, estrogens and androgens (Levin, 2013). According to the American Cancer Society's guidelines on nutrition and physical activity for cancer prevention, weight control, eating practices and physical activity are second only to tobacco use as modifiable determinants of cancer risk (Kushi et al., 2012). Weight loss, weight management and weight-gain prevention through diet and physical activity are among the leading modifiable determinants of cancer risk, in part because cancer risk increases linearly with adult weight gain (American Cancer Society, 2013a, 2013b; Schubart, Stuckey, Ganeshamoorthy, & Sciamanna, 2011).

During the past decade, while obesity rates have plateaued for women, there has been a significant positive linear trend in obesity among men, with African American men increasing at one of the fastest rates (Flegal, Carroll, Kit, & Ogden, 2012; Ogden, Carroll, Kit, & Flegal, 2012) and almost two-fifths (38.8%) of African American men aged 20 years and older classified as obese (Flegal, Carroll, Ogden, & Curtin, 2010; Flegal et al., 2012; Ogden, et al., 2012). African American men have higher rates of developing and dying from cancers associated with obesity, poor diet and physical inactivity than White men, White women and African American women (Griffith, Metzl, & Gunter, 2011; Ward et al., 2004). It is projected that by 2030 obesity will lead to about 500,000 additional cases of cancer in the U.S. (National Cancer Institute, 2012); however, if every adult reduced their BMI by 1% or 1 kg, about 100,000 new cases of cancer might be avoided (National Cancer Institute, 2012). With this rise in obesity, there is an urgent need to develop weight loss interventions to promote healthier and more active lifestyles for African American men (American Cancer Society, 2013a).

2.2.1 Importance of caloric intake and physical activity

Diets rich in fruits and vegetables have an inverse association with the risk of developing leading cancers (Campbell et al., 2007; Davidson et al., 2013; Sigurdardottir et al., 2012), and regular physical activity can reduce the risks of colon, prostate, and lung cancers (Friedenreich & Orenstein, 2002; Kruk & Czerniak, 2013; Thune & Furberg, 2001). Rates of fruit and vegetable consumption among African American men are consistently lower than any other racial/ethnic group of men or women (Wolf et al., 2008). Furthermore, African American men are more likely than White men to have diets that are high in fat, sodium, red meat, calories and sugar-sweetened beverages (Airhihenbuwa et al., 1996; Bleich, Wang, Wang, & Gortmaker, 2009; James, 2004). Diets high in animal fat, especially fat from red meat, are associated with higher risk of prostate and colorectal cancer among men (Gillum & Griffith, 2010; Williams et al., 2011). Encouraging healthy eating through increased intake of fruit and vegetables and decreased intake of red meat can be especially beneficial to reducing the risk of cancers for African American men.

Throughout the life course, African American men's rates of physical activity tend to decline from adolescence into young adulthood and are lowest during middle age. Men are typically most active during childhood and adolescence but their levels of physical activity decrease in young adulthood (18–29 years), followed by stable low levels in middle adulthood (30–64 years) (Caspersen, Pereira, & Curran, 2000; Griffith, Ellis, & Allen, 2013). African American men under 35 years of age report engaging in physical activity through competitive team sports and athletics, while middle and older aged African American men who remain active tend to do so through informal and often solitary activities such as household chores, yard work, walking, and biking (Wanko et al., 2004; Wood, 2002).

African American men over age 45 are also less likely than their White counterparts to participate in leisure time physical activity (25% vs. 36%) or walking for exercise (43% vs. 52%) (Brownson, Boehmer, & Luke, 2005). Efforts to explain the modifiable factors that could reduce these differences and improve activity rates overall have been limited. Researchers have found that middle-aged African American men often prefer physical activities in a social atmosphere, such as involving family, the community, or a partner or buddy (Hooker, Wilcox, Rheaume, Burroughs, & Friedman, 2011; Parham & Scarinci, 2007) but no such studies have examined the activity preferences of young African American men.

A qualitative study of 49 middle-aged and older African American men identified a number of factors that seemed to influence physical activity (Friedman, Hooker, Wilcox, Burroughs, & Rheaume, 2012; Hooker, Harmon, Burroughs, Rheaume, & Wilcox, 2011; Hooker, Wilcox, Burroughs, Rheaume, & Courtenay, 2012; Hooker, Wilcox, et al., 2011). Barriers discussed in this study included: lack of time, limited access to places to engage in physical activity, inadequate social support, lack of motivation, physical ailments and chronic health conditions. Noted facilitators of physical activity included: receiving positive messages about physical activity from a trustworthy and reliable source; making physical activity enjoyable; physical activity interventions with peer social interaction, social support, and competition; and having spousal support but only limited spousal involvement. Obesity, unhealthy eating and physical inactivity are important, modifiable determinants of African American men's cancer risk.

2.3 What do we know about promoting healthy eating, physical activity or weight loss in African American men?

Reviews of physical activity or physical fitness intervention studies with African American adults (Whitt-Glover & Kumanyika, 2009) and reviews of faith-based physical activity interventions (Bopp, Peterson, & Webb, 2012) found that few African American men were included and only one study focused exclusively on men (Whitt-Glover & Kumanyika, 2009). Behavioral interventions targeted to African Americans typically included fewer than 30% men (Bopp et al., 2006; Griffith, Metzl, & Gunter, 2011; Newton, Griffith, Kearney, & Bennett, 2014; Warren et al., 2010; Whitt-Glover & Kumanyika, 2009), highlighting the need to develop strategies to increase the participation of African American men in healthy caloric intake, physical activity and weight loss interventions. In addition to the limited literature on African American men and physical activity, recent reviews of physical activity in men that focused on worksites, walking groups and physical activity found only two studies reporting data for African American men (George et al., 2012; Kassavou, Turner, & French, 2013; Wong, Gilson, van Uffelen, & Brown, 2012). To address this gap in the literature, Newton and colleagues conducted a review of studies published before January 2013 to determine the current state of the evidence concerning African American men's response to interventions to promote weight loss or increase healthy eating or physical activity (Newton et al., 2014). This systematic review found only six community-based studies that reported data separately for African American men (Newton et al., 2014). African American men were an exclusive sample in four studies, including one that focused on healthy eating (Wolf, Lepore, Vandergrift, Basch, & Yaroch, 2009) and one that focused on weight loss (Treadwell et al., 2010). While the study by Treadwell et al. found a 2.54 kg weight loss in African American men who were obese or overweight, it was a small scale study that included only 42 men, and the study lacked a control group and only lasted 6 weeks (i.e., it had a high risk of bias). Since this review, the findings from two additional community-based physical activity interventions have been published—*Men on the Move and Men on the Move- Nashville*. These studies similarly produced modest yet significant increases in African American men's perceived health status, self-efficacy to sustain physical activity and intensity of physical activity (Dean, Griffith, McKissic, Cornish, & Johnson-Lawrence, 2018; Griffith, Allen, Johnson-Lawrence, & Langford, 2014). *Men on the Move-Nashville* was the first study to our knowledge that used SMS text messaging and physical activity tracking devices to enhance the effectiveness of a physical activity intervention for African American men (Kirwan, Duncah, Vandelanotte, & Mummery, 2012).

While a few noteworthy studies are currently in the field (Active and Healthy Brotherhood [PCORI]; *Mighty Men* [American Cancer Society]; TAILOR MADE: Solutions for your Health [National Institute on Minority Health and Health Disparities, NIMHD]), no rigorously designed community-based, behavioral interventions to promote physical activity or weight loss have been completed and there has only been one rigorously designed healthy eating intervention (Wolf et al., 2009) for African American men (Griffith, Bergner, Cornish, & McQueen, 2018).

2.4 Why did we choose to conduct our weight loss study of middle-aged and older African American men in faith-based organizations?

One of the biggest challenges in promoting the health of African American men is identifying settings that are not only appropriate to conduct an intervention but that will allow the intervention to potentially be sustained after the study has been completed. Given limited viable options for settings to recruit as well as conduct and sustain interventions with African American men, faith-based organizations are important sites to consider because of their standing and ubiquity in the African American community (Bruce & Whitt-Glover, 2013). In our work, we have found that this is particularly true for recruiting middle-aged and older African American men (Griffith, Jaeger, Sherman, & Moore, 2019). We will briefly review selected literature and discuss our experience recruiting African American men to participate in research studies through different settings (i.e., fraternal organizations, barbershops) and conclude with an explanation for how we chose faith-based organizations to reach our specific group of African American men.

2.4.1 Fraternal organizations as a site to conduct research with African American men

In prior formative research, we attempted to develop a weight loss intervention for African American men through men's civic, social and fraternal organizations (Griffith, Gunter, & Allen, 2012). Whether founded to support African Americans in urban areas or to provide social and educational support to aspiring college and graduate students on college campuses, fraternal organizations have been an important institution within African American communities and particularly among African American men. The African American Greek letter organizations that were originally founded for college students, quickly grew to also support African American men who have graduated and moved on to a new phase of life. The stated purpose and mission of the organization when The National Pan-Hellenic Council, Incorporated (NPHC) was founded in 1930 was to bring "Unanimity of thought and action as far as possible in the conduct of Greek letter collegiate fraternities and sororities, and to consider problems of mutual interest to its member organizations." The NPHC is currently composed of nine (9) International Greek letter Sororities and Fraternities, but because of our focus on men's health, we focused on the fraternities: Alpha Phi Alpha Fraternity, Inc., Iota Phi Theta Fraternity, Inc., Kappa Alpha Psi Fraternity, Inc., Phi Beta Sigma Fraternity, Inc. and Omega Psi Phi Fraternity, Inc. These fraternal and civic organizations have national and international graduate chapters, connecting men of African descent around the world beyond their college years. These organizations serve as spaces to help convene and support African American men, and are potential settings to intervene to improve African American men's eating and physical activity. In these settings, we aim to build on the values, support and camaraderie of organizations that are influential in the lives of African American men.

In our experience, fraternities present some challenges for intervention activities and may not be the ideal organizational contexts for interventions to improve African American men's health, but this may not be true of other communities or contexts. Fraternities typically meet on a monthly basis. Thus, the frequency of contact among members is not regular enough to utilize those meetings as a way to build accountability and social support. When members gather outside of these meetings, frequently the gatherings are focused on either social activities that are not necessarily health-promoting or on service activities to help others. Rarely do members of these organizations gather to focus on their own health, though there is beginning to be a shift in this culture. For example, Omega Psi Phi Fraternity, Inc. has collaborated with NIMHD to launch *Brother, You're on My Mind: Changing the National Dialogue Regarding Mental Health among African American Men*. Additionally, the confidential nature of membership to these organizations poses barriers for intervention activities or community presentations and does not allow for an open setting for intervention by people who are not members of the organization. In addition to civic and fraternal organizations, we also considered barbershops.

2.4.2 Barbershops as a site to conduct research with African American men

Barbershops are important institutions for African American men. In addition to grooming services, barbershops often provide a safe space for men to connect and communicate about issues most salient to them (Cowart, Brown, & Biro, 2004; Hart & Smith, 2008; Hood, Hall, Dixon, Jolly, & Linnan, 2017; Releford, Frencher, & Yancey, 2010). The barbershop is known as a place for frank, unfettered dialog, which some suggest facilitates their ability to conduct successful health promotion interventions in this setting. Some researchers have trained barbers to be peer educators, because of their trusted role in the community, but others have suggested they could be much more than that (Releford et al., 2010). These spaces represent one of the few trusted places where researchers have been consistently able to access and engage African American men, particularly to recruit participants or educate them about various initiatives.

Murphy et al. (2017) surveyed 127 African American men in 25 Chicago area barbershops to investigate locations, community areas and settings to engage African American men in research and health promotion. While successful health interventions have been recruited for and disseminated through barbershops (Linnan et al., 2011), Murphy and colleagues found that the top recommended recruitment sites varied by age. The sites these men identified were: 18–29 year old city park or a recreational center; 30–39 year old gym, bars or the street; 40–49 year old various stores, especially home improvement stores, and the mall; and 50-year old fast food restaurants in the mornings (e.g., McDonald's) and individuals' homes. The authors found that it is important to consider the heterogeneity among this population by age and sexual orientation in identifying appropriate locations to recruit African American men. This information encouraged us to explore alternative sites for reaching men and for influencing their health behavior, in an attempt to explore a variety of organizations that are central to the lives and relationships of African American men.

Our experience, however, is that while barbershops are a great venue to do outreach and health education with this population, they are a difficult place to conduct and sustain an intervention where health education needs to take place. Barbershops are a great venue to reach men to share brief educational information, recruit participants for surveys and studies, or conduct screenings. It is currently unclear, however, how equipped these settings are to support weekly small groups that take more than a few minutes or that would need to be sustained for several months to years.

2.4.3 Faith-based organizations as a site to recruit African American men to participate in research

Despite fewer African American men than African American women attending faith-based organizations on a weekly basis, African American faith-based organizations still reach more men than any other organization or institution in the U.S. (Bopp et al., 2012; Gillum, 2005; Gillum & Ingram, 2006; Levin, 2013). Religion and spirituality often are important aspects of men's lives and help facilitate the attainment of a healthier lifestyle (Bopp et al., 2012; Garfield, Isacco, & Sahker, 2013). Faith-based organizations have often been trusted places where African American men go for a safe and affirming environment to address health and social issues with peers (Gwin, Taylor, Branscum, & Hofford, 2013). Historically, African American faith-based organizations have been a trusted source of social support and a means of providing health services and education (Billingsley, 1999; Bopp, Baruth, Peterson, & Webb, 2013; Bopp et al., 2012; Bruce & Whitt-Glover, 2013; Campbell et al., 2007). Community-based interventions aimed at improving health and reducing health disparities help people to increase control over and improve their individual and collective health by emphasizing strategies to empower local communities to alter their environments to their advantage (Griffith, Pichon, Campbell, & Allen, 2010; Guldan, 1996). Semmes (1996) defines the process of making these structural changes to make a community healthier as creating an *institutionalized health ethic*, or a process of coordinating community residents, health professionals, health institutions, and health resources to increase awareness and change norms to make a community healthier. While fraternal organizations, barbershops and other community-based settings and institutions have been successful in helping to recruit African American men, arguably none of these interventions have greater capacity to institutionalize, or build on an institutionalized health ethic, than faith-based organizations, even for African American men. Thus, there have been calls for more faith-based interventions for men (Bopp et al., 2012; Bruce & Whitt-Glover, 2013), but there are no published, rigorously designed faith-based healthy eating, physical activity, or weight loss interventions for African American men (Newton et al., 2014). *Mighty Men* strives to address this gap. Next, we describe the approach, conceptual foundations and proposed mechanisms that underlie *Mighty Men*.

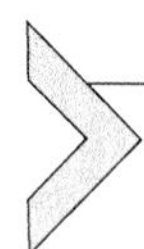

3. Conceptual and theoretical foundations of mighty men

Mighty Men is a faith-based weight loss intervention to reduce cancer disparities. Francis and colleagues suggest that faith-based health promotion programs include five key components: (1) involve the faith community in all aspects of the research process; (2) recognize the time constraints faced by the senior pastor or faith leader; (3) incorporate spirituality and religious beliefs into prevention messages in a way that is non-judgmental and allows for some flexibility; (4) ensure cultural appropriateness; and (5) create a sense of ownership by the faith-based organization (Francis & Liverpool, 2009). We have designed *Mighty Men* with these factors in mind and structured the program to be consistent with these goals. These elements of *Mighty Men* are consistent with principles of community-based participatory research (Viswanathan et al., 2004): acknowledge the African American faith community as a unique unit of identity; build on the strengths and resources of the African American faith community; foster capacity building and mutual learning of all community and academic partners; balance knowledge generation and the intervention; share decision-making power; begin with a topic identified by faith leaders; focus on the local relevance of the problem; and disseminate results to the community. The intervention model is anchored in population factors that shape the social determinants which affect the health of people who are African American, male and middle-aged or older (Griffith, 2015; Griffith, Ellis, et al., 2013; McLeroy, Bibeau, Steckler, & Glanz, 1988) (see Fig. 1). In our conceptual model, the social environment describes the cultural values, beliefs, and norms regarding diverse aspects of life (including but not limited to health), as well as social institutions, networks, and resources in these settings. Though we are not measuring the impact of weight loss on cancer risk, these factors highlight key determinants that influence weight loss and cancer risk.

A minority health promotion approach (NIMHD, 2017) encouraged us to develop tailored interventions based upon the health determinant findings for specific population groups, in this case, focusing exclusively on African American men to identify cultural strengths and barriers that may serve as

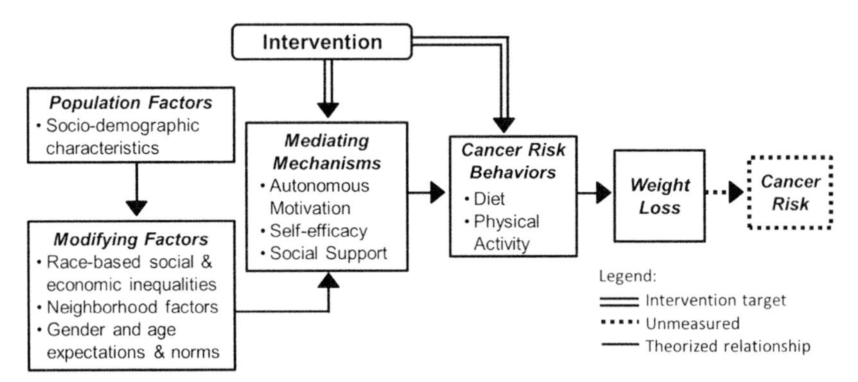

Fig. 1 Conceptual framework of intervention.

important components of interventions to improve their health. For African American men, gender, race, and class interact in complex ways to increase the risk for various illnesses including cancer (Griffith, 2012, 2015, 2016). These socio-demographic characteristics represent key aspects of African American men's lives that are fundamental to understanding the challenges of promoting health in this population (Griffith, Gunter, & Watkins, 2012; Xanthos, Treadwell, & Holden, 2010). Jackson and Knight identify that stressful social and economic living conditions, combined with restricted access to potential resources to manage those conditions, contribute to behavioral responses to stress that may adversely affect health outcomes (Jackson & Knight, 2006; Mezuk et al., 2013).

We use a Social Ecological Framework (McLeroy et al., 1988) of health behavior to highlight modifying factors that are not influenced by the intervention, but that can independently affect the outcomes and mediating mechanisms, such as those that define the obesogenic environment (Hill, 2006; Hill, Wyatt, Reed, & Peters, 2003). We do not seek to change these modifying factors in this intervention but we know that we have to help men develop the motivation, skills and social resources to overcome these environmental constraints (Fortier, Williams, Sweet, & Patrick, 2009; Resnicow & McMaster, 2012). Race-based social and economic inequalities shape the neighborhood factors where African American men live, and the resources they have or lack to promote healthy eating and physical activity (e.g., full service grocery stores, limited safe places to exercise) (Griffith, Cornish, McKissic, & Dean, 2016; Griffith & Johnson, 2012; Williams, 2003; Williams & Collins, 2001; Xanthos et al., 2010). Gender ideology and perceived gender norms highlight the fact that despite significant changes in sex roles, gender norms and gendered expectations (Griffith, Gilbert, Bruce, & Thorpe, 2016), fulfilling traditionally masculine roles is a core aspect of how African American men define themselves and are defined by peers, family and community members (Griffith & Cornish, 2018). These gendered notions have been found to compete with men's efforts to engage in healthy behavior and maintain a healthy lifestyle (Griffith, Ellis, et al., 2013; Griffith, et al., 2011; Griffith, Gunter, & Allen, 2012; Griffith & Johnson, 2012). *Mighty Men* focuses on the benefits of healthy eating and physical activity in helping to fulfill key social and family roles, not simply because it is good for men's health, but because, as our work has shown, African American men define manhood as the ability to fulfill these roles (Griffith, 2015; Griffith & Cornish, 2018). Health often matters to African American men and is prioritized because it affords them the opportunity to achieve social roles that are important to them and that affect other people in their families and social networks (Allen, Griffith, & Gaines, 2013; Griffith, Brinkley-Rubinstein, Bruce, Thorpe, & Metzl, 2015; Griffith, Ellis, et al., 2013).

Mediating mechanisms are the primary psychological vehicles through which *Mighty Men* is designed to affect African American men's cancer risk behaviors (e.g., healthy eating and physical activity), weight, and cancer risk. Our formative qualitative research and pilot studies reinforce the notion that autonomous motivation, self-efficacy and social support are key mediators of African American men's healthy eating and physical activity (Allen et al., 2013; Griffith et al., 2014; Griffith, Ellis, & Allen, 2012; Griffith, et al., 2011; Griffith, King, & Allen, 2013; Griffith, Wooley, & Allen, 2013).

Mighty Men is designed to enhance autonomous motivation (Fortier et al., 2009), which is a key element of Self-Determination Theory (Ryan & Deci, 2000; Teixeira, Silva, Mata, Palmeira, & Markland, 2012; Vansteenkiste, Niemiec, & Soenens, 2010). Behaviors that are autonomous result from conscious choices, are personally relevant, and are not performed due to pressure or coercion either by external or internal forces such as pleasing others, fear of disease, or avoiding guilt, anxiety, shame or consequences. Autonomous motivation links health and health behavior with individual African American men's broader values and goals and increases the likelihood that they will consciously choose to increase and sustain healthy behaviors. Autonomous motivation is key because it is the primary factor that helps men overcome modifying factors that constrain opportunities and motivation to engage in healthier behaviors. The autonomously motivated individuals not only see the importance of the behavior, but also connect change with their core values and beliefs. These individuals feel competent, are ready to take action and persist when faced with obstacles and have identified meaningful reasons for change (Resnicow, Teixeira, & Williams, 2017).

In addition to autonomous motivation, the second key mediating mechanism is to increase self-efficacy, or the confidence one has in his ability to exert control over his own motivation, behavior, and social environment (Bandura, 2004). Social Cognitive Theory (Bandura, 2004) posits that self-efficacy affects health behavior directly and indirectly through its impact on goals, outcome expectations, and perceived facilitators and impediments. Thus, an important objective of health interventions is to increase African American men's perceived ability to engage in healthier behavior despite environmental, social and time pressures that may present barriers to healthy eating or physical activity.

The intervention design and setting provide opportunities for observational learning and social support (Bandura, 2004). Emotional, instrumental, informational, and appraisal social support (Anderson, Winett, & Wojcik, 2007; House, 1981; Israel & McLeroy, 1985) are critical targets of our intervention, since gendered norms and expectations are an important part of influencing men's health behaviors (Creighton & Oliffe, 2010; Evans et al., 2011; Mróz, Chapman, Oliffe, & Bottorff, 2011), and there is a need to help men rethink how they balance health behavior and social roles. Next, we will describe the components of *Mighty Men*.

3.1 Mighty men intervention components and rationale

Mighty Men is a 6-month weight loss intervention that includes four components: (a) tailored goals/messages (via SMS text message), (b) self-monitoring (via wearable device (Fitbit) and SMS text message), (c) small group training and education and (d) educational and community-based information and resources. These intervention components were selected based on our formative research (Dean, et al., 2018; Griffith et al., 2014) and experience using them in prior studies and implemented simultaneously as they complement one another (see Table 1) (Vandelanotte, Reeves, Brug, & De Bourdeaudhuij, 2008). While all of these intervention components have not been tested together in an intervention for this population, the development of *Mighty Men* was informed by variations and enhancements of Resnicow's *Body and Soul*, *Healthy Body Healthy Spirit* and *Eat For Life* tailoring interventions (Resnicow et al., 2004, 2008, 2005, 2002)

Table 1 Relationship between intervention components and theoretical constructs.

Intervention component	Preliminary studies references	Conceptual target
Tailored goals and messages	Tailoring HE and PA, (Alexander et al., 2010; Bennett et al., 2012; Davis et al., 2010; Foley et al., 2012; Greaney et al., 2009; Resnicow et al., 2008)	Skills training motivation
Self-monitoring	MOTM Nashville, (Bennett et al., 2012; Foley et al., 2012; Greaney et al., 2009)	Self-efficacy motivation
Small group training/ meetings	MOTM Michigan (Griffith et al., 2014), MOTM Nashville	Skills training knowledge motivation social support
Educational and community-based information and resources	MOTM Nashville, (Bennett et al., 2012; Foley et al., 2012; Greaney et al., 2009; Resnicow et al., 2008)	Knowledge feedback motivation

and Bennett's *Be Fit, Be Well* (BFBW) intervention (Greaney et al., 2009). The small group training and education component was added after the success of Griffith's *Men on the Move* pilot intervention with African American men (Dean, et al., 2018; Griffith et al., 2014). We found that the small group component of the intervention facilitated social support and motivation by connecting spirituality, values and life goals with healthy eating and physical activity (Cornish, McKissic, Dean, & Griffith, 2017).

3.1.1 Tailored goals and messages

Since its introduction in the early 1990s, hundreds of studies have shown that materials and messages tailored to individual needs, preferences, and psychological profiles increase recipient attention, processing effort, and perceived message salience, thereby outperforming generic or group-targeted messages in promoting health behavior change. Tailored strategies can heighten the personal relevance of intervention content, while minimizing cost and labor intensity (Hawkins, Kreuter, Resnicow, Fishbein, & Dijkstra, 2008; Kreuter, Strecher, & Glassman, 1999). Over the last decade, we have worked to develop the framework of the *Mighty Men* intervention to individually tailor health communications for weight loss to African American men. This collaborative approach to tailoring is unique in that it marries two approaches to individual tailoring. The *Mighty Men* approach to tailoring combines Dr. Ken Resnicow's previous work on tailoring to ethnic identity, core values, goals, and motivation (Davis et al., 2010; Davis & Resnicow, 2012; Resnicow, Baranowski, Ahluwalia, & Braithwaite, 1999; Resnicow et al., 2009) with Dr. Gary Bennett's efforts to tailor to an algorithm of behavior- change goals that produce weight loss (Greaney et al., 2009). The approach to tailoring to African American men developed specifically for *Mighty Men* uses manhood (Griffith, 2015; Griffith et al., 2015; Griffith, Pennings, Bruce, & Ayers, 2019) as the core identity (Griffith & Cornish, 2018) that links men's values, goals, and motivation to engage in health-promoting behavior. Tailoring to manhood incorporates the heterogeneity among African American men's identities and definitions of health.

Each week, one individually-tailored motivational SMS message is sent to participants to connect the goal of being healthier by engaging in healthier behavior to positive ways to achieve a goal, aspiration or value with notions of manhood (Fortier et al., 2009; Resnicow & McMaster, 2012; Ryan & Deci, 2000; Teixeira et al., 2012). These motivational messages are based on a novel conceptualization of African American manhood (Griffith & Cornish, 2018) because we hypothesize that this framing of the message will increase attention to and effortful processing of the messages, since the messages are particularly salient and relevant to their values, goals, identities and lives (Hawkins et al., 2008). Manhood refers to an age-appropriate social status and aspirational identity that reflects the embodiment of virtuous characteristics and traits, performance of certain social roles, and the fulfillment of gendered expectations associated with being an adult male (Griffith, 2015; Vandello & Bosson, 2013). Our formative research has shown that men tend to define manhood and health by their ability to successfully achieve life goals and fulfill social and professional roles (Griffith et al., 2015). For middle-aged and older African American men, we found that adhering to busy work schedules, striving for success in their careers, seeking to fulfill the role of financial provider, balancing family time and responsibilities, and being active in their community were critical priorities that were congruent with their phase of life and more valued than making time to engage in healthy eating or physical activity (Diemer, 2002; Griffith, et al., 2011). Our tailored motivational messages help counter these norms by helping men to recognize and identify ways to create and maintain a healthy lifestyle *and* fulfill these important roles.

In addition to motivational messages tailored to manhood, *Mighty Men* participants focus on four main behavior change goals designed to produce an energy deficit. These goals are based on the *interactive Obesity Treatment Approach (iOTA)*, which was explicitly designed to identify goals that induce an energy deficit; the iOTA approach was created by Dr. Gary Bennett, founder and director of the Duke Global Digital Health Science Center, and he has successfully used this system in several intervention trials (Bennett & Glasgow, 2009; Bennett et al., 2010, 2012; Foley et al., 2012; Greaney et al., 2009). The iOTA system utilizes a new media interface (e.g., web, interactive voice response, text messaging) to self-monitor participant adherence to a set of simple obesogenic behavior change goals that have been selected based on their empirical support and relevance to the target population (Greaney et al., 2009). During the baseline and mid-point study visit, each *Mighty Men* participant completes a short risk behavior assessment. Based on this assessment, a computer algorithm selects four behavior change goals, prioritizing those behaviors: (1) in highest need of change; (2) for which the participant has high self-efficacy and readiness for change; (3) for which the participant identifies few change barriers, and; (4) that fulfill the intended caloric deficit (Greaney et al., 2009). These goals may change every 3 months to maintain novelty and minimize habituation.

3.1.2 Self-monitoring

Regular self-monitoring is a robust predictor of weight loss, although adherence typically wanes over time. Self-monitoring fatigue likely results from its cognitive complexity and lack of immediate feedback. These challenges are magnified among those with lower literacy levels.

Data from previous studies show that new media self-monitoring tools can be effective with African Americans (Bennett & Glasgow, 2009; Bennett et al., 2010, 2012; Foley et al., 2012; Greaney et al., 2009). *Mighty Men* uses telephone technologies for several reasons.

First, SMS text messaging helps overcome the literacy/numeracy barriers associated with paper logs. Second, these systems have high reach, as mobile phone penetration is very high among African American men (Smith, 2010). According to the Pew Research Center, rates of internet use among African Americans (80%) are slightly lower than White Americans (87%), but smartphone use rates among African Americans ages 18–29 (85%) are higher than their White counterparts (79%) (Smith, 2014). Young adults and lower income African Americans are often dependent on smartphones for internet use. This literature also reveals that among African American men, 12% do not use broadband service at home, but own a smartphone; thus, the majority of this group's internet use comes from their smartphone. Additionally, 96% of African Americans ages 18–29 use a form of social networking (i.e., Facebook, LinkedIn or Twitter) at least once a day. Lastly, these technologies are inexpensive to develop, are immediately scalable, and thus are easily utilized to promote health among African American men.

3.1.3 Small group training/meetings

Mighty Men includes weekly small group meetings led by a certified personal trainer and/or a facilitator. These sessions last no more than an hour and a half and are designed to increase men's motivation, knowledge and skills and foster healthy support and competition. The rationale for this is based on wanting the men to focus initially on the interrelationship between self-efficacy, autonomous motivation and enjoyment (Fortier et al., 2009). We believe that one of the reasons people may lose motivation to maintain healthy patterns of eating and physical activity is that they simply no longer enjoy it nor do they see the benefit directly related to things that are most important to them. Consequently, the small group sessions are designed to help integrate skills building (and therefore self-efficacy); personal values, spirituality and goal setting to facilitate autonomous motivation; and enjoyment of physical activity and healthy eating. In focus groups conducted as part of the process evaluation of Men on the Move- Nashville, the men noted that they enjoyed the camaraderie and support they got from their small group, and that they benefitted from seeing that others were struggling with and overcoming similar barriers to physical activity (Dean, et al., 2018). The men reported that they learned from and supported one another with strategies to overcome these barriers to physical activity.

Connecting healthy eating and physical activity to spirituality, personal values and goals further reinforces the advantage of conducting *Mighty Men* in faith-based organizations. The social network, doctrine and resources of the intervention and faith-based organization can coalesce into a sustainable way for the faith-based organization to connect with men and for men to find autonomous motivation to maintain healthier lifestyles.

Since time pressures and schedules were identified as key barriers to physical activity (Griffith, et al., 2011), it was important to have men identify a regular meeting time that would fit their schedules, which is a lesson we learned in Men on the Move (Griffith et al., 2014). The weekly sessions consist of 45 min of a small group discussion followed by 45 min of physical activity. The small group aspects of the intervention are modeled on motivational interviewing principles and the success of previous group-based motivational interviewing, albeit with women and around HIV (Holstad, DiIorio, Kelley, Resnicow, & Sharma, 2011). In addition to the physical activity during the small group sessions, the men are encouraged to engage in any physical activity throughout the week that gets them moving and to focus on identifying types of physical activity that they enjoy.

3.1.4 Educational and community-based information and resources

In order to increase the competence and confidence of men who participate in *Mighty Men* so that the changes they make throughout the duration of the program are sustainable, they receive updates about community events, resources, and information that is supportive of healthy lifestyles. This provides them with tools and resources to identify and engage in other community-based health initiatives as well as increases their knowledge of available community resources.

4. Discussion and next steps

In this chapter, we described the rationale for *Mighty Men: A Faith-Based Weight Loss Intervention for African American Men*. Using a minority health promotion approach (NIMHD, 2017), we illustrate how focusing exclusively on African American men facilitates identifying cultural strengths that could be the foundation of interventions to improve their health (Bediako & Griffith, 2007; Griffith, 2016; Griffith & Cornish, 2018; Griffith, Metzl, & Gunter, 2011; Jack & Griffith, 2013). Some men find pride and new opportunities that come with aging and accompanying bodily changes, but others experience a sense of impending crisis along with increasing awareness of their own mortality (Evans et al., 2011; Griffith, Cornish, Bergner, Bruce, & Beech, 2017). Regardless, health has meaning that we have to consider outside of the narrow confines of anthropometric tests, biological processes and genetic risk factors. We have to align health interventions with the definitions of health and well-being that are held by members of the population of interest if we hope to be optimally successful (Griffith et al., 2015; Ravenell, Johnson, & Whitaker, 2006; Robertson, 2006).

One of the key points we highlight in this chapter is the importance of the setting where we may intervene to reach and sustain interventions for middle-aged and older African American men. Our formative research also found that some African American men root their conceptions and definitions of manhood in their faith (Griffith & Cornish, 2018). More broadly, faith-based organizations historically have been an important moral, social, and cultural foundation of many aspects of the African American community and its informational and social support networks (Billingsley, 1999; Bruce & Whitt-Glover, 2013; Griffith, Pichon, et al., 2010). The Black church has long been considered a critical institution in the African American community (Taylor, Chatters, & Levin, 2004), as it is one of the few stable and independent Black institutions with the means and resources to organize people (Billingsley, 1999). Faith-based organizations have often assumed responsibility for filling the health and human service needs abdicated or inadequately and unacceptably met by other service institutions that serve or are part of the African American community (Billingsley, 1999; Bruce & Whitt-Glover, 2013; Griffith, Pichon, et al., 2010; Levin, 2013). Churches have previously been involved in health interventions in a variety of ways (Peterson, Atwood, & Yates, 2002), and are an important vehicle for disseminating accurate information in a sensitive and culturally appropriate manner (Coyne-Beasley & Schoenbach, 2000; Francis & Liverpool, 2009; Griffith, Campbell, et al., 2010). However, faith-based organizations are underutilized as tools to promote the health of African American men.

Despite the demographic shifts in who attends church and the declining role that faith-based organizations may have in the U.S. and the African American community specifically (Levin, 2013), no other institution has replaced the faith community as a way to reliably reach, intervene and institutionalize an intervention to promote health and well-being among African American adults. *Mighty Men* highlights the importance of utilizing an intersectional approach to understand the unique ways that we approach creating programs, policies and institutions to promote the health and well-being of men who are African American, adult, male, and of a certain age or phase of life. These combine in important ways that are relevant for determining the setting to house an intervention, vehicles to disseminate information, modes of communication and focus and goals of the research (Griffith, Gunter, et al., 2012; Griffith, Metzl, & Gunter, 2011). We built our intervention on strengths rather than the negative images and narratives that plague discussions of African American men and African American men's health in a way such that we strive to be consistent with the work of Dr. Collins Airhihenbuwa, a pioneer in developing culturally-grounded interventions. In his words, "...when you arrive in a community to address a health issue, you should begin with something positive that the community does correctly—their assets. If you cannot identify something positive, then you should not remain in the community, otherwise you are likely to focus only on their problems and may in fact become a part of their problems" (Airhihenbuwa, 2010). We built *Mighty Men* on the strengths of the African American community and African American men and we look forward to testing its effects and seeing its outcomes.

Acknowledgments

This paper has been supported in part by Vanderbilt University, the American Cancer Society (RSG-15-223-01-CPPB), and NIH/NIMHD (5U54MD010722-02).

References

Airhihenbuwa, C. O. (2010). Culture matters in global health. *European Health Psychologist, 12*(4), 52–55.

Airhihenbuwa, C. O., Kumanyika, S., Agurs, T. D., Lowe, A., Saunders, D., & Morssink, C. B. (1996). Cultural aspects of African American eating patterns. *Ethnicity & Health, 1*(3), 245–260.

Alexander, G. L., McClure, J. B., Calvi, J. H., Divine, G. W., Stopponi, M. A., Rolnick, S. J., et al. (2010). A randomized clinical trial evaluating online interventions to improve fruit and vegetable consumption. *American Journal of Public Health, 100*(2), 319–326. https://doi.org/10.2105/ajph.2008.154468.

Allen, J. O., Griffith, D. M., & Gaines, H. C. (2013). "She looks out for the meals, period": African American men's perceptions of how their wives influence their eating behavior and dietary health. *Health Psychology, 32*(4), 447–455.

American Cancer Society. (2009). *Cancer facts & figures for African Americans 2009–2010.* Retrieved from https://www.cancer.org/content/dam/cancer-org/research/cancer-facts. and-statistics/cancer-facts-and-figures-for-african-americans/cancer-facts-and-figures. for-african-americans-2009-2010.pdf.

American Cancer Society. (2013a). *Body weight and cancer risk.* Retrieved from http://www.cancer.org/cancer/cancercauses/dietandphysicalactivity/bodyweightandcancer risk/body-weight-and-cancer-toc.

American Cancer Society. (2013b). *Cancer facts & figures for African Americans 2013–2014.* Retrieved from https://www.cancer.org/content/dam/cancer-org/research/cancer-facts. and-statistics/cancer-facts-and-figures-for-african-americans/cancer-facts-and-figures. for-african-americans-2013-2014.pdf.

Anderson, E. S., Winett, R. A., & Wojcik, J. R. (2007). Self-regulation, self-efficacy, outcome expectations, and social support: Social cognitive theory and nutrition behavior. *Annals of Behavioral Medicine, 34*(3), 304–312.

Bandura, A. (2004). Health promotion by social cognitive means. *Health Education & Behaviour, 31*(2), 143–164.

Bediako, S. M., & Griffith, D. M. (2007). Eliminating racial/ethnic health disparities: Reconsidering comparative approaches. *Journal of Health Disparities Research and Practice, 2*(1), 49–62.

Bennett, G. G., & Glasgow, R. E. (2009). The delivery of public health interventions via the internet: Actualizing their potential. *Annual Review of Public Health, 30,* 273–292.

Bennett, G. G., Herring, S. J., Puleo, E., Stein, E. K., Emmons, K. M., & Gillman, M. W. (2010). Factors associated with physical activity among African-American men and women. *American Journal of Preventive Medicine, 30*(4), 340–346.

Bennett, G., Warner, E., Glasgow, R., Askew, S., Goldman, J., Ritzwoller, D., et al. (2012). Obesity treatment for socioeconomically disadvantaged patients in primary care practice. *JAMA Internal Medicine, 172*(7), 565–574.

Billingsley, A. (1999). *Mighty like a river: The black church and social reform*. New York: Oxford University Press.

Bleich, S. N., Wang, Y. C., Wang, Y., & Gortmaker, S. L. (2009). Increasing consumption of sugar-sweetened beverages among US adults: 1988-1994 to 1999-2004. *The American Journal of Clinical Nutrition, 89*(1), 372–381.

Bopp, M. P., Baruth, M. P., Peterson, J. A., & Webb, B. L. (2013). Leading their flocks to health? Clergy health and the role of clergy in faith-based health promotion interventions. *Family & Community Health, 36*(3), 182–192.

Bopp, M., Peterson, J. A., & Webb, B. L. (2012). A comprehensive review of faith-based physical activity interventions. *American Journal of Lifestyle Medicine, 6*(6), 460–478.

Bopp, M., Wilcox, S., Laken, M., Butler, K., Carter, R. E., McClorin, L., et al. (2006). Factors associated with physical activity among African-American men and women. *American Journal of Preventive Medicine, 30*(4), 340–346.

Brownson, R. C., Boehmer, T. K., & Luke, D. A. (2005). Declining rates of physical activity in the United States: What are the contributors? *Annual Review of Public Health, 26*(1), 421–443.

Bruce, M. A., & Whitt-Glover, M. C. (2013). Faith-based initiatives to promote health. *Family & Community Health, 36*(3), 179–181.

Campbell, M. K., Hudson, M. A., Resnicow, K., Blakeney, N., Paxton, A., & Baskin, M. (2007). Church-based health promotion interventions: Evidence and lessons learned. *Annual Review of Public Health, 28*, 213–234.

Caspersen, C. J., Pereira, M. A., & Curran, K. M. (2000). Changes in physical activity patterns in the United States, by sex and cross-sectional age. *Medicine and Science in Sports Exercise, 32*(9), 1601–1609.

Centers for Disease Control and Prevention. (2016). *Defining adult overweight and obesity*. Retrieved from https://www.cdc.gov/obesity/adult/defining.html.

Cornish, E. K., McKissic, S., Dean, D., & Griffith, D. M. (2017). Lessons learned about motivation from a pilot physical activity intervention for African American men. *Health Promotion Practice, 18*(1), 102–109.

Courtenay, W. H. (2000). Constructions of masculinity and their influence on men's well-being: A theory of gender and health. *Social Science and Medicine, 50*(10), 1385–1401.

Courtenay, W. H. (2002). A global perspective on the field of men's health: An editorial. *International Journal of Men's Health, 1*(1), 1.

Cowart, L. W., Brown, B., & Biro, D. J. (2004). Educating African American men about prostate cancer: The barbershop program. *American Journal of Health Studies, 19*(4), 205.

Coyne-Beasley, T., & Schoenbach, V. J. (2000). The African-American church: A potential forum for adolescent comprehensive sexuality education. *Journal of Adolescent Health, 26*(4), 289–294.

Creighton, G., & Oliffe, J. L. (2010). Theorising masculinities and men's health: A brief history with a view to practice. *Health Sociology Review, 19*(4), 409–418.

Davidson, E. M., Liu, J. J., Bhopal, R. A. J., White, M., Johnson, M. R. D., Netto, G., et al. (2013). Behavior change interventions to improve the health of racial and ethnic minority populations: A tool kit of adaptation approaches. *Milbank Quarterly, 91*(4), 811–851.

Davis, R., Alexander, G., Calvi, J., Wiese, C., Greene, S., Nowak, M., et al. (2010). A new audience segmentation tool for African Americans: The black identity classification scale. *Journal of Health Communication, 15*(5), 532.

Davis, R. E., & Resnicow, K. (2012). The cultural variance framework for tailoring health messages. *Health communication message design: Theory and practice*, 115–136.

Dean, D., Griffith, D. M., McKissic, S., Cornish, E. K., & Johnson-Lawrence, V. (2018). Men on the Move—Nashville: Feasibility and acceptability of a technology-enhanced, physical activity pilot intervention for overweight and obese, middle and older-age African American men. *American Journal of Men's Health, 12*(4), 798–811.

Diemer, M. A. (2002). Constructions of provider role identity among African American men: An exploratory study. *Cultural Diversity and Ethnic Minority Psychology, 8*(1), 30–40.

Evans, J., Frank, B., Oliffe, J. L., & Gregory, D. (2011). Health, Illness, Men and Masculinities (HIMM): A theoretical framework for understanding men and their health. *Journal of Men's Health, 8*(1), 7–15.

Flegal, K., Carroll, M., Kit, B., & Ogden, C. (2012). Prevalence of obesity and trends in the distribution of Body Mass Index among US adults, 1999-2010. *JAMA, 307*(5), 491–497.

Flegal, K. M., Carroll, M. D., Ogden, C. L., & Curtin, L. R. (2010). Prevalence and trends in obesity among US adults, 1999-2008. *Journal of the American Medical Association, 303*(3), 235–241.

Foley, P., Levine, E., Askew, S., Puleo, E., Whiteley, J., Batch, B., et al. (2012). Weight gain prevention among black women in the rural community health center setting: The Shape Program. *BMC Public Health, 12*(1), 305.

Fortier, M. S., Williams, G. C., Sweet, S. N., & Patrick, H. (2009). Self-determination theory: Process models for health behavior change. In R. J. DiClemente, R. A. Crosby, & M. Kegler (Eds.), *Vol. 2. Emerging theories in health promotion practice and research: Strategies for improving public health* (pp. 157–183). Jossey-Bass.

Francis, S. A., & Liverpool, J. (2009). A review of faith-based HIV prevention programs. *Journal of Religion and Health, 48*(1), 6–15.

Friedenreich, C. M., & Orenstein, M. R. (2002). Physical activity and cancer prevention: Etiologic evidence and biological mechanisms. *The Journal of Nutrition, 132*(11), 3456S–3464S.

Friedman, D. B., Hooker, S. P., Wilcox, S., Burroughs, E. L., & Rheaume, C. E. (2012). African American men's perspectives on promoting physical activity: "We're not that difficult to figure out!" *Journal of Health Communication, 17*(10), 1151–1170.

Garfield, C. F., Isacco, A., & Sahker, E. (2013). Religion and spirituality as important components of men's health and wellness an analytic review. *Journal of Psychology and Theology*, 7(3), 27–37.

George, E., Kolt, G., Duncan, M., Caperchione, C., Mummery, W. K., Vandelanotte, C., et al. (2012). A review of the effectiveness of physical activity interventions for adult males. *Sports Medicine, 42*(4), 281–300.

Gillum, R. F. (2005). Frequency of attendance at religious services and cigarette smoking in American women and men: The Third National Health and Nutrition Examination Survey. *Preventive Medicine, 41*(2), 607–613.

Gillum, F., & Griffith, D. M. (2010). Prayer and spiritual practices for health reasons among American adults: The role of race and ethnicity. *Journal of Religion and Health, 49*(3), 283–295.

Gillum, R. F., & Ingram, D. D. (2006). Frequency of attendance at religious services, hypertension, and blood pressure: The third national health and nutrition examination survey. *Psychosomatic Medicine, 68*(3), 382–385.

Greaney, M., Quintiliani, L., Warner, E., King, D., Emmons, K., Colditz, G., et al. (2009). Weight management among patients at community health centers: The "Be Fit, Be Well" study. *Obesity and Weight Management, 5*(5), 222–228.

Griffith, D. M. (2012). An intersectional approach to men's health. *Journal of Men's Health, 9*(2), 106–112.

Griffith, D. M. (2015). "I AM a Man": Manhood, minority men's health and health equity. *Ethnicity and Disease, 25*(3), 287–293.

Griffith, D. M. (2016). Biopsychosocial approaches to men's health disparities research and policy. *Behavioral Medicine, 42*(3), 211–215.

Griffith, D., Allen, J., DeLoney, E., Robinson, K., Lewis, E., Campbell, B., et al. (2010). Community-based organizational capacity building as a strategy to reduce racial health disparities. *The Journal of Primary Prevention, 31*(1), 31–39.

Griffith, D. M., Allen, J. O., Johnson-Lawrence, V., & Langford, A. (2014). Men on the move: A pilot program to increase physical activity among African American men. *Health Education & Behavior, 41*(2), 164–172.

Griffith, D. M., Bergner, E., Cornish, E. K., & McQueen, C. M. (2018). Physical activity interventions with African American or Latino men: A systematic review. *American Journal of Men's Health, 12*(4), 1102–1117.

Griffith, D. M., Brinkley-Rubinstein, L., Bruce, M. A., Thorpe, R. J., Jr., & Metzl, J. M. (2015). The interdependence of African American men's definitions of manhood and health. *Family and Community Health, 38*(4), 284–296.

Griffith, D. M., Campbell, B., Allen, J. O., Robinson, K. J., & Stewart, S. K. (2010). YOUR Blessed Health: An HIV-prevention program bridging faith and public health communities. *Public Health Reports, 125*(Suppl. 1), 4.

Griffith, D. M., & Cornish, E. K. (2018). "What defines a man?": Perspectives of African American men on the components and consequences of manhood. *Psychology of Men & Masculinity, 19*(1), 78–88.

Griffith, D. M., Cornish, E. K., Bergner, E. M., Bruce, M. A., & Beech, B. M. (2017). "Health is the ability to manage yourself without help": How older African American men define health and successful aging. *The Journals of Gerontology Series B: Psychological Sciences and Social Sciences, 73*(2), 240–247.

Griffith, D. M., Cornish, E. K., McKissic, S. A., & Dean, D. A. L. (2016). Differences in perceptions of the food environment between African American men who did and did not consume recommended levels of fruits and vegetables. *Health Education & Behavior, 43*(6), 648–655.

Griffith, D. M., Ellis, K. R., & Allen, J. O. (2012). How does health information influence African American men's health behavior? *American Journal of Men's Health, 6*(2), 156–163.

Griffith, D. M., Ellis, K. R., & Allen, J. O. (2013). Intersectional approach to stress and coping among African American men. *American Journal of Men's Health,* 7(4S), 16–27.

Griffith, D. M., Gilbert, K. L., Bruce, M. A., & Thorpe, R. J., Jr. (2016). Masculinity in men's health: Barrier or portal to healthcare? In *Men's health in primary care* (pp. 19–31). Springer.

Griffith, D. M., Gunter, K., & Allen, J. O. (2011). Male gender role strain as a barrier to African American men's physical activity. *Health Education & Behavior, 38*(5), 482–491. https://doi.org/10.1177/1090198110383660.

Griffith, D. M., Gunter, K., & Allen, J. O. (2012). A systematic approach to developing contextual, culturally, and gender sensitive interventions for African American men: The example of men 4 health. In R. Elk & H. Landrine (Eds.), *Cancer disparities: Causes and evidence-based solutions* (pp. 193–210). NY: Springer Publishing.

Griffith, D. M., Gunter, K., & Watkins, D. C. (2012). Measuring masculinity in research on men of color: Findings and future directions. *American Journal of Public Health, 102*(Suppl. 2), S187–S194. https://doi.org/10.2105/AJPH.2012.300715.

Griffith, D. M., Jaeger, E. C., Sherman, L. D., & Moore, H. J. (2019). Middle-aged men's health: Patterns and causes of health disparities during a pivotal period in the lifecourse. In D. M. Griffith, M. A. Bruce, & R. J. Thorpe, Jr., (Eds.), *Men's health equity.* New York, NY: Routledge.

Griffith, D. M., & Johnson, J. L. (2012). Implications of racism for African American men's cancer risk, morbidity and mortality. In H. M. Treadwell, C. Xanthos, & K. B. Holden (Eds.), *Social determinants of health among African-American men* (pp. 21–38). San Francisco: Wiley.

Griffith, D. M., King, A., & Allen, J. O. (2013). Male peer influence on African American men's motivation for physical activity: Men's and women's perspectives. *American Journal of Men's Health*, 7(2), 169–178.

Griffith, D. M., Mason, M., Rodela, M., Matthews, D. D., Tran, A., Royster, M., et al. (2007). A structural approach to examining prostate cancer risk for rural, southern African American men. *Journal of Health Care for the Poor and Underserved*, *18*(Suppl), 73–101.

Griffith, D. M., Metzl, J. M., & Gunter, K. (2011). Considering intersections of race and gender in interventions that address U.S. men's health disparities. *Public Health*, *125*(7), 417–423.

Griffith, D. M., Pennings, J. S., Bruce, M. A., & Ayers, G. D. (2019). Measuring the dimensions of African American Manhood: A factor analysis. In D. M. Griffith, M. A. Bruce, & R. J. Thorpe, Jr., (Eds.), *Men's health equity*. New York, NY: Routledge.

Griffith, D. M., Pichon, L. C., Campbell, B., & Allen, J. O. (2010). YOUR Blessed Health: A faith-based CBPR approach to addressing HIV/AIDS among African Americans. *AIDS Education and Prevention*, *22*(3), 203–217.

Griffith, D. M., Wooley, A. M., & Allen, J. O. (2013). "I'm ready to eat and grab whatever I can get" determinants and patterns of African American men's eating practices. *Health Promotion Practice*, *14*(2), 181–188.

Guldan, G. S. (1996). Obstacles to community health promotion. *Social Science & Medicine*, *43*(5), 689–695.

Gwin, S. M. S., Taylor, E. L. P., Branscum, P. P. R. D., & Hofford, C. P. (2013). Assessment of factors that predict physical activity among Oklahoma clergy: A theory of planned behavior approach. *Family & Community Health July/September*, *36*(3), 193–203.

Hart, A., Jr., & Smith, W. R. (2008). Recruiting African-American barbershops for prostate cancer education. *Journal of the National Medical Association*, *100*(9), 1012.

Hawkins, R. P., Kreuter, M., Resnicow, K., Fishbein, M., & Dijkstra, A. (2008). Understanding tailoring in communicating about health. *Health Education Research*, *23*(3), 454.

Hebert, J. R., Hurley, T. G., Harmon, B. E., Heiney, S., Hebert, C. J., & Steck, S. E. (2012). A diet, physical activity, and stress reduction intervention in men with rising prostate-specific antigen after treatment for prostate cancer. *Cancer Epidemiology*, *36*(2), e128–e136.

Hill, J. O. (2006). Understanding and addressing the epidemic of obesity: An energy balance perspective. *Endocrine Reviews*, *27*(7), 750–761. https://doi.org/10.1210/er.2006-0032.

Hill, J., Wyatt, H., Reed, G., & Peters, J. (2003). Obesity and the environment: Where do we go from here? *Science*, *299*(5608), 853–855.

Holstad, M. M., DiIorio, C., Kelley, M. E., Resnicow, K., & Sharma, S. (2011). Group motivational interviewing to promote adherence to antiretroviral medications and risk reduction behaviors in HIV infected women. *AIDS and Behavior*, *15*(5), 885–896.

Hood, S., Hall, M., Dixon, C., Jolly, D., & Linnan, L. (2017). Organizational-level recruitment of barbershops as health promotion intervention study sites: Addressing health disparities among black men. *Health Promotion Practice*, *19*(3), 377–389.

Hooker, S. P., Harmon, B., Burroughs, E. L., Rheaume, C. E., & Wilcox, S. (2011). Exploring the feasibility of a physical activity intervention for midlife African American men. *Health Education Research*, *26*(4), 732–738.

Hooker, S. P., Wilcox, S., Burroughs, E. L., Rheaume, C. E., & Courtenay, W. (2012). The potential influence of masculine identity on health-improving behavior in midlife and older African American men. Journal of Men's Health, 9(2), 79–88.

Hooker, S. P., Wilcox, S., Rheaume, C. E., Burroughs, E. L., & Friedman, D. B. (2011). Factors related to physical activity and recommended intervention strategies as told by midlife and older African American men. *Ethnicity and Disease, 21*(3), 261–267.

House, J. S. (1981). *Work stress and social support*. Reading, MA: Addison-Wesley Educational Publishers, Inc.

Israel, B. A., & McLeroy, K. R. (1985). Social networks and social support: Implications for health education. Introduction. *Health Education Quarterly, 12*(1), 1–4.

Jack, L., & Griffith, D. M. (2013). The health of African American men: Implications for research and practice. *American Journal of Men's Health*, 7(4 Suppl), 5S–7S.

Jackson, J. S., & Knight, K. M. (2006). Race and self-regulatory health behaviors: The role of the stress response and the HPA axis. In K. W. Schaie & L. L. Carstensten (Eds.), *Social structure, aging and self-regulation in the elderly* (pp. 189–240). New York: Springer.

James, D. C. S. (2004). Factors influencing food choices, dietary intake, and nutrition-related attitudes among African Americans: Application of a culturally sensitive model. *Ethnicity & Health, 9*(4), 349–367.

Kassavou, A., Turner, A., & French, D. P. (2013). Do interventions to promote walking in groups increase physical activity? A meta-analysis. *International Journal of Behavioral Nutrition and Physical Activity, 10*(1), 18.

Key, T., Schatzkin, A., Willett, W., Allen, N., Spencer, E., & Travis, R. (2004). Diet, nutrition and the prevention of cancer. *Public Health Nutrition*, 7(1a), 187–200.

Kirwan, M., Duncah, M. J., Vandelanotte, C., & Mummery, W. K. (2012). Using smartphone technology to monitor physical activity in the 10,000 steps program: A matched case-control trial. *Journal of Medical Internet Research, 14*(2), e55.

Koh, H. K., Massin-Short, S., & Elqura, L. (2009). Disparities in tobacco use and lung cancer. In H. K. Koh (Ed.), *Toward the elimination of cancer disparities* (pp. 109–135). New York: Springer.

Kreuter, M. W., Strecher, V. J., & Glassman, B. (1999). One size does not fit all: The case for tailoring print materials. *Annals of Behavioral Medicine, 21*(4), 276.

Kruk, J., & Czerniak, U. (2013). Physical activity and its relation to cancer risk: Updating the evidence. *Asian Pacific Journal of Cancer Prevention, 14*, 3993–4003.

Kushi, L. H., Doyle, C., McCullough, M., Rock, C. L., Demark-Wahnefried, W., Bandera, E. V., et al. (2012). American Cancer Society guidelines on nutrition and physical activity for cancer prevention. *CA: A Cancer Journal for Clinicians, 62*(1), 30–67.

Levin, J. (2013). Faith-based initiatives in health promotion: History, challenges, and current partnerships. *American Journal of Health Promotion, 28*(3), 139–141.

Linnan, L. A., Reiter, P. L., Duffy, C., Hales, D., Ward, D. S., & Viera, A. J. (2011). Assessing and promoting physical activity in African American barbershops: Results of the FITStop pilot study. *American Journal of Men's Health, 5*(1), 38–46.

McLeroy, K. R., Bibeau, D., Steckler, A., & Glanz, K. (1988). An ecological perspective on health promotion programs. *Health Education Quarterly, 15*(4), 351–377.

Mezuk, B., Abdou, C. M., Hudson, D., Kershaw, K. N., Rafferty, J. A., Lee, H., et al. (2013). "White Box" epidemiology and the social neuroscience of health behaviors: The environmental affordances model. *Society and Mental Health, 3*(2), 79–95.

Mróz, L. W., Chapman, G. E., Oliffe, J. L., & Bottorff, J. L. (2011). Men, food, and prostate cancer: Gender influences on men's diets. *American Journal of Men's Health, 5*(2), 177–187.

Murphy, A. B., Moore, N. J., Wright, M., Gipson, J., Keeter, M., Cornelious, T., et al. (2017). Alternative locales for the health promotion of African American men: A survey of African American men in Chicago barbershops. *Journal of Community Health, 42*(1), 139–146.

Najari, B. B., Rink, M., Li, P. S., Karakiewicz, P. I., Scherr, D. S., Shabsigh, R., et al. (2012). Sex disparities in cancer mortality: The risks of being a man in the United States. *The Journal of Urology, 189*(4), 1470–1474.

National Cancer Institute. (2012). *Fact sheet: Obesity and cancer risk*. Retrieved from www.cancer.gov website http://www.cancer.gov/cancertopics/factsheet/Risk/obesity/print.

Newton, R. L., Jr., Griffith, D. M., Kearney, W., & Bennett, G. G. (2014). A systematic review of physical activity, dietary, and weight loss interventions involving African American men. *Obesity Reviews, 15*(Suppl. S4), 93–106.

NIMHD. (2017). *National institute on minority health and health disparities: justification of budget request*. Retrieved from https://www.nimhd.nih.gov/docs/congress- justification/2018CJ.pdf.

Ogden, C. L., Carroll, M. D., Kit, B. K., & Flegal, K. M. (2012). Prevalence of obesity among adults: United States. *NCHS Data Brief, 2013*(131), 1–8.

Parham, G. P., & Scarinci, I. C. (2007). Strategies for achieving healthy energy balance among African Americans in the Mississippi Delta. *Preventing Chronic Disease, 4*(4), A97. doi:A97 [pii].

Peterson, A. (2009). Future research agenda in men's health. In A. B. P. Tovey (Ed.), *Men's health: Body, identity and social context* (pp. 202–213). West Sussex, UK: Wiley-Blackwell.

Peterson, J., Atwood, J. R., & Yates, B. (2002). Key elements for church-based health promotion programs: Outcome-based literature review. *Public Health Nursing, 19*(6), 401–411.

Ravenell, J. E., Johnson, W. E., & Whitaker, E. E. (2006). African American men's perceptions of health: A focus group study. *Journal of the National Medical Association, 98*, 544–550.

Releford, B. J., Frencher, S. K., Jr., & Yancey, A. K. (2010). Health promotion in barbershops: Balancing outreach and research in African American communities. *Ethnicity & Disease, 20*(2), 185.

Resnicow, K., Baranowski, T., Ahluwalia, J. S., & Braithwaite, R. L. (1999). Cultural sensitivity in public health: Defined and demystified. *Ethnicity & Disease, 9*(1), 10–21.

Resnicow, K., Campbell, M. K., Carr, C., McCarty, F., Wang, T., Periasamy, S., et al. (2004). Body and soul. A dietary intervention conducted through African-American churches. *American Journal of Preventive Medicine, 27*(2), 97–105.

Resnicow, K., Davis, R., Zhang, G., Konkel, J., Strecher, V., Shaikh, A., et al. (2008). Tailoring a fruit and vegetable intervention on novel motivational constructs: Results of a randomized study. *Annals of Behavioral Medicine, 35*(2), 159–169.

Resnicow, K., Davis, R., Zhang, N., Strecher, V., Tolsma, D., Calvi, J., et al. (2009). Tailoring a fruit and vegetable intervention on ethnic identity: Results of a randomized study. *Health Psychology, 28*(4), 394.

Resnicow, K., Jackson, A., Blissett, D., Wang, T., McCarty, F., Rahotep, S., et al. (2005). Results of the healthy body healthy spirit trial. *Health Psychology, 24*(4), 339–348.

Resnicow, K., Jackson, A., Braithwaite, R., DiIorio, C., Blisset, D., Rahotep, S., et al. (2002). Healthy body/healthy spirit: A church-based nutrition and physical activity intervention. *Health Education Research, 17*(5), 562–573.

Resnicow, K., & McMaster, F. (2012). Motivational interviewing: Moving from why to how with autonomy support. *International Journal of Behavioral Nutrition and Physical Activity, 9*(1), 19.

Resnicow, K., Teixeira, P. J., & Williams, G. C. (2017). Efficient allocation of public health and behavior change resources: The "Difficulty by Motivation" matrix. *American Journal of Public Health, 107*(1), 55–57.

Robertson, S. (2006). 'I've been like a coiled spring this last week': Embodied masculinity and health. *Sociology of Health and Illness, 28*(4), 433–456.

Robertson, S. (2007). *Understanding men and health: Masculinities, identity, and well-being*. Maidenhead: Open University Press.

Ryan, R. M., & Deci, E. L. (2000). Self-determination theory and the facilitation of intrinsic motivation, social development, and well-being. *American Psychologist, 55*(1), 68–78.

Schubart, J. R., Stuckey, H. L., Ganeshamoorthy, A., & Sciamanna, C. N. (2011). Chronic health conditions and internet behavioral interventions: A review of factors to enhance user engagement. *Computers Informatics Nursing, 29*(2), 81–92.

Semmes, C. E. (1996). *Racism, health, and post-industrialism: A theory of African-American health*. Westport, CT: Praeger.

Sigurdardottir, L. G., Valdimarsdottir, U. A., Fall, K., Rider, J. R., Lockley, S. W., Schernhammer, E., et al. (2012). Circadian disruption, sleep loss, and prostate cancer risk: A systematic review of epidemiologic studies. *Cancer Epidemiology Biomarkers & Prevention, 21*(7), 1002–1011.

Smith, A. (2010). *Technology trends among people of color*. Pew Internet & American Life Project. Retrieved from.

Smith, A. (2014). *African Americans and technology use: A demographic portrait*. Washington, DC: Pew Research Center.

Taylor, R., Chatters, L., & Levin, J. (2004). Impact of religion on mental health and well-being. In *Religion in the lives of African-Americans: Social, psychological, and health perspectives*. Thousand Oaks, CA: Sage Publications, Inc.

Taylor, K. L., Turner, R. O., Davis, J. L., 3rd, Johnson, L., Schwartz, M. D., Kerner, J., et al. (2001). Improving knowledge of the prostate cancer screening dilemma among African American men: An academic-community partnership in Washington, DC. *Public Health Reports, 116*(6), 590–598.

Teixeira, P. J., Silva, M. N., Mata, J., Palmeira, A. L., & Markland, D. (2012). Motivation, self- determination, and long-term weight control. *International Journal of Behavioral Nutrition and Physical Activity, 9*(1), 22.

Thune, I., & Furberg, A.-S. (2001). Physical activity and cancer risk: Dose-response and cancer, all sites and site-specific. *Medicine and Science in Sports and Exercise, 33*(6 Suppl), S530–S550. discussion S609-510.

Treadwell, H., Holden, K., Hubbard, R., Harper, F., Wright, F., Ferrer, M., et al. (2010). Addressing obesity and diabetes among African American men: Examination of a community-based model of prevention. *Journal of the National Medical Association, 102*(9), 794.

Vandelanotte, C., Reeves, M. M., Brug, J., & De Bourdeaudhuij, I. (2008). A randomized trial of sequential and simultaneous multiple behavior change interventions for physical activity and fat intake. *Preventive Medicine, 46*(3), 232.

Vandello, J. A., & Bosson, J. K. (2013). Hard won and easily lost: A review and synthesis of theory and research on precarious manhood. *Psychology of Men & Masculinity, 14*(2), 101–113.

Vansteenkiste, M., Niemiec, C. P., & Soenens, B. (2010). The development of the five mini-theories of self-determination theory: An historical overview, emerging trends, and future directions. *Advances in Motivation and Achievement, 16*, 105–165.

Viswanathan, M., Ammerman, A., Eng, E., Gartlehner, G., Lohr, K. N., Griffith, D. M., et al. (2004). *Community-based participatory research: Assessing the evidence*. Rockville, MD: Agency for Healthcare Research and Quality.

Wanko, N. S., Brazier, C. W., Young-Rogers, D., Dunbar, V. G., Boyd, B., George, C. D., et al. (2004). Exercise preferences and barriers in urban African Americans with type 2 diabetes. *Diabetes Education, 30*(3), 502–513.

Ward, E., Jemal, A., Cokkinides, V., Singh, G. K., Cardinez, C., Ghafoor, A., et al. (2004). Cancer disparities by race/ethnicity and socioeconomic status. *CA: A Cancer Journal for Clinicians, 54*(2), 78.

Warren, T. Y., Barry, V., Hooker, S. P., Sui, X., Church, T. S., & Blair, S. N. (2010). Sedentary behaviors increase risk of cardiovascular disease mortality in men. *Medicine and Science in Sports Exercise, 42*(5), 879–885.

Wei, E. K., Giovannucci, E., Wu, K., Rosner, B., Fuchs, C. S., Willett, W. C., et al. (2004). Comparison of risk factors for colon and rectal cancer. *International Journal of Cancer, 108*(3), 433.

Whitt-Glover, M. C., & Kumanyika, S. K. (2009). Systematic review of interventions to increase physical activity and physical fitness in African-Americans. *American Journal of Health Promotion, 23*(6), S33–S56.

Williams, D. R. (2003). The health of men: Structured inequalities and opportunities. *American Journal of Public Health, 93*(5), 724–731.

Williams, D. R., & Collins, C. (2001). Racial residential segregation: A fundamental cause of racial disparities in health. *Public Health Reports, 116*(5), 404–416.

Williams, T. T., Griffith, D. M., Pichon, L. C., Campbell, B., Allen, J. O., & Sanchez, J. C. (2011). Involving faith-based organizations in adolescent HIV prevention. *Progress in Community Health Partnerships: Research, Education, and Action, 5*(4), 425–431.

Wolf, R. L., Lepore, S. J., Vandergrift, J. L., Basch, C. E., & Yaroch, A. L. (2009). Tailored telephone education to promote awareness and adoption of fruit and vegetable recommendations among urban and mostly immigrant black men: A randomized controlled trial. *Preventive Medicine, 48*(1), 32–38.

Wolf, R. L., Lepore, S. J., Vandergrift, J. L., Wetmore-Arkader, L., McGinty, E., Pietrzak, G., et al. (2008). Knowledge, barriers, and stage of change as correlates of fruit and vegetable consumption among urban and mostly immigrant Black men. *Journal of the American Dietetic Association, 108*(8), 1315–1322.

Wong, J. Y. L., Gilson, N. D., van Uffelen, J. G. Z., & Brown, W. J. (2012). The effects of workplace physical activity interventions in men: A systematic review. *American Journal of Men's Health, 6*(4), 303–313.

Wood, F. G. (2002). Ethnic differences in exercise among adults with diabetes. *Western Journal of Nursing Research, 24*(5), 502–515.

Xanthos, C., Treadwell, H. M., & Holden, K. B. (2010). Social determinants of health among African American men. *Journal of Men's Health, 7*(1), 11–19.

CHAPTER TEN

A primer for cancer research programs on defining and evaluating the catchment area and evaluating minority clinical trials recruitment

Lynne H. Nguyen[a,*], Elise D. Cook[b]

[a]Department of Health Disparities Research, The University of Texas MD Anderson Cancer Center, Houston, TX, United States

[b]Department of Clinical Cancer Prevention, The University of Texas MD Anderson Cancer Center, Houston, TX, United States

*Corresponding author: e-mail address: lhnguyen@mdanderson.org

Contents

Abstract

National Cancer Institute (NCI) designated cancer centers are charged with reducing disparities, improving cancer-related health outcomes, and increasing clinical trial participation for the catchment area population. Succeeding in this endeavor requires a clear definition of each cancer center's geographic catchment area as well as the demographic characteristics of the populations residing in the catchment area. For this reason, the definition of the catchment area is now a required element of NCI grant

Advances in Cancer Research, Volume 146
ISSN 0065-230X
https://doi.org/10.1016/bs.acr.2020.02.001

applications. This primer provides detailed information related to the definition of cancer centers' catchment areas and provides a case example from the University of Texas MD Anderson Cancer Center to highlight best practice strategies for compiling and interpreting cancer health statistics for the catchment area.

1. Introduction

National Cancer Institute (NCI) designated cancer centers are expected to reduce the overall cancer burden in their geographic area (also referred to as an institution's "catchment area" (Tai & Hiatt, 2017)). In addition, major clinical research institutes are expected to include sufficient representation of minority populations and to properly report race, ethnicity, and sex (Clayton & Tannenbaum, 2016; Griggs, Maingi, Blinder, et al., 2017; Polite, Adams-Campbell, Brawley, et al., 2017). There is often confusion among cancer researchers and program reviewers about how to properly define the catchment area and evaluate how well patients accrued to clinical trials reflect the demographics in their catchment area. In this review, a NCI comprehensive cancer center, the University of Texas MD Anderson Cancer Center (MD Anderson), is used to illustrate best practices in this realm. The data collected from MD Anderson's catchment area, the state of Texas, are presented.

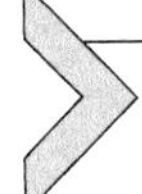

2. Methods

2.1 Requirements for a justifiable catchment area definition

Defining a catchment area allows a hospital to describe its primary patient population. Only with a clear definition and understanding of the catchment area can one determine how well the institution meets the needs of its primary patient population. The NCI states that the catchment area must be defined and justified by the center based on the geographic area it serves. The catchment area must be population based, using census tracts, zip codes, county or state lines or geographically defined boundaries, and it must include the local area surrounding the cancer center. For a research center that has not determined a catchment area, some considerations for defining the catchment area include: a fixed distance about the facility; a fixed travel time around the facility; a fixed proportion of patients in a geographic area;

a predetermined geographic area based on your mission and vision, or a combination of the above. Other factors that may impact on how the catchment area is defined includes natural geographic boundaries that affect who seek care at your site, the presence and strength of competitors, research ongoing in certain geographical areas, and the expertise of the center's researchers.

2.2 Determine the demographics of the catchment area

The next step is to gain an understanding of the catchment area demographics, since these variables can affect cancer incidence. Age is the most significant risk factor for cancer, and the cancer population typically reflects the demographics of older adults in the catchment area. Accrual goals for treatment studies should match the demographics of the cancer patient population in your catchment area. In addition, race, ethnicity, and sex can impact cancer rates of a population. Finally, population size and immigration patterns are important to consider. A population with a high cancer rate, but a small population size means a smaller total number of patients, and thus a smaller pool of potential clinical trial participants. In general, immigrants/emigrants to an area tend to be younger, and thus have lower cancer risk. Current demographic information is key to the assessment of our constantly changing catchment areas.

2.3 Data required to assess the catchment area

Listed below is a description of data that will be needed to properly assess a given catchment area.

- Demographics of the catchment area (ex: race, ethnicity, sex, age, zip code)
- Cancers that affect residents of the catchment area (review absolute number of cases as well as incidence and mortality rates, and stage of disease at diagnosis), as well as cancer risk factors
- Demographics of cancer patients in the community and any cancer/health disparities that disproportionately affect specific populations in the catchment area (review incidence and mortality rates by race, ethnicity, sex, and stage of diagnosis)
- Demographics of patients treated in your center and the types of cancer treated
- Demographics of patients enrolled in clinical trials

2.4 Available sources to extract the required data

Sources for these data may include your state cancer registry, state data center for population data, US census data, The Surveillance, Epidemiology, and End Results (SEER) program, hospital tumor registry data, and data available from the American College of Surgeons related to cancer type by state and treatment facility type among other variables.

2.5 Evaluate the institution's reach and service provided to its catchment area

With the catchment area defined and understood, the final step is to evaluate how well your facility equitably serves the catchment area. Two steps to help you assess your center's reach and service:

2.5.1 Assess your total patient base by race, ethnicity, and sex

- What proportion of patients are from the catchment area?
- What proportion are from outside the catchment area?
- How are the demographics of these two groups of patients similar, or dissimilar? What factors impact your patient composition?
- Do the proportion of patients in your center that come from the catchment area reflect the proportion of people diagnosed with cancer in the catchment area? i.e., do the patients inside the hospital look like the patients in the community?

2.5.2 Assess the patients enrolled on your clinical studies

The clinical studies comparators for NCI-designated cancer centers are those on therapeutic interventional trials; those on nontherapeutic interventional trials; and those on noninterventional trials (e.g., epidemiologic, observational trials or biobanking studies)

- What percentage of patients are enrolled on a clinical study? (if they came to your center, did they participate on a study?)
- Do the demographics of clinical trial participants reflect the demographics of your patient population? i.e., did everyone have equal access to research?

3. Results

3.1 MD Anderson's overall patient population and catchment area demographics

Texas is a "minority–majority" state, meaning that minority populations comprised a majority (57%) while non-Hispanic Whites accounted for

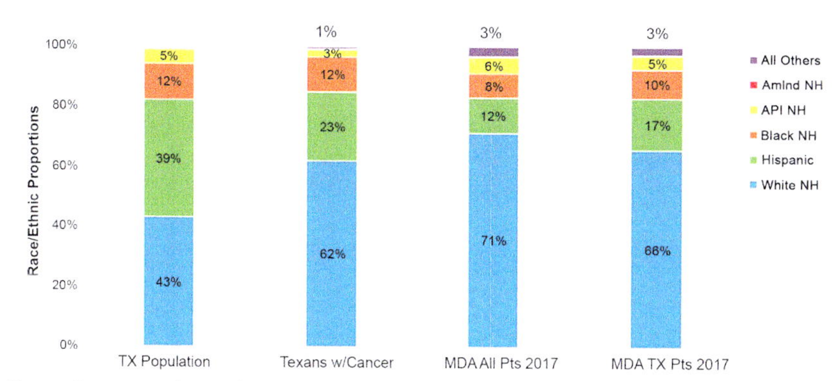

Fig. 1 Demographics of Population and Cancer in Texas and at MD Anderson.

43% of the population. [Fig. 1.] Confoundingly, new cancer Texas incidence data showed a reversed scenario, with minorities comprising only 38% of all new cancer cases, while non-Hispanic Whites comprised 62% of new cases. Cancer data have been provided by the Texas Cancer Registry (2020). Two concurrent phenomenon contributed to this reversed scenario: age and population demographics. Older age is a significant risk factor for cancer, so the cancer population tends to resemble the demographics of older people. In Texas, the majority of people aged 65 and older (61%) in 2016 were non-Hispanic White.

At first glance, comparing MD Anderson new patients with cancer to newly diagnosed Texans would show a patient population that poorly reflects the catchment area (71% of MD Anderson patients were non-Hispanic White, compared to 62% of newly diagnosed Texans.) However, a closer review of patients' zip codes showed that a substantial proportion of MD Anderson patients (44%) came from outside of our catchment area of Texas. Patients from outside Texas were predominantly non-Hispanic White, and reflected the cancer demographics of the US. In order to assess whether our patients from Texas were representative of the Texas cancer population, we needed to compare similar populations; the demographics of our Texas patients with the demographics of all new cancer cases in Texas. The last bar in Fig. 1 showed that our patients from Texas more closely reflect the demographics of Texans diagnosed with cancer.

3.2 Access of patients in MD Anderson's catchment area to clinical trials

It is important to compare clinical trial participants with all patients who received cancer care at an institution, in order to determine whether all

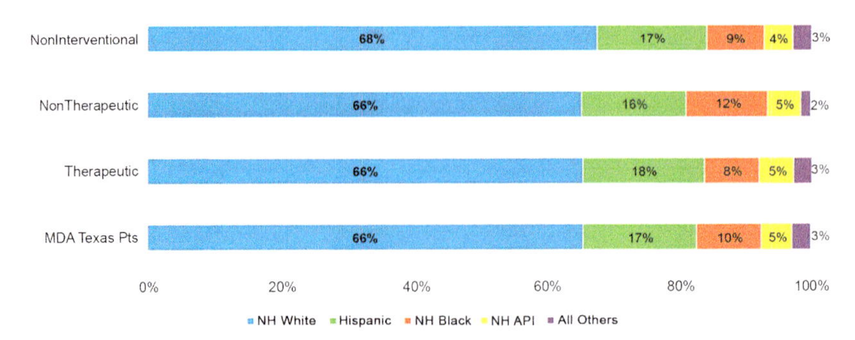

Fig. 2 MD Anderson race/ethnicity of patients enrolled on trials.

cancer patients have an equitable opportunity to participate on a trial. Fig. 2 below illustrates MD Anderson Cancer Center FY2017 data. The proportions of Hispanic and Black patients enrolled on therapeutic and nontherapeutic trials are very similar to the proportion of Texas patients treated at the Center that are Hispanic, Black, or Asian/Pacific Islander.

4. Discussion

When a particular race, ethnic group, or sex has higher rates of some cancers, the demographic characteristics of the patient population may skew accordingly, depending on the size of the specific population at risk. For example, although the average Texan is younger than the average US resident, the proportion of Texans aged 65 and older is growing quickly. By 2042, 1 in 5 Texans will be 65 or older. Thus, one can expect the number of newly diagnosed cancer cases to continue to increase, even as cancer rates drop, largely due to the impact of the population size of the older segment, as well as aging of the population. Even as the population of aged Texans continues to grow, the demographics of those older also evolve. By 2022, Whites will comprise less than 50% of those aged 65 and older. Cancer centers and physicians will see a significant shift in patient demographics, with a greater proportion being non-white. Furthermore, some population groups may have higher cancer burden for some cancers. For example, in Texas, with an incidence rate of 13.5/100,000, Hispanics have the highest burden of liver cancer among all race/ethnic groups. This rate is almost double the rate for Non-Hispanic Whites (7.0/100,000) (SEER, 2014).

The goals for providing adequate reach and service to the institution's catchment area are (1) That the Cancer Center's patient population reflect the population of people diagnosed with the condition in the catchment area and (2) that the patients on clinical studies reflect the patients who seek care at the Cancer Center, indicating that all patients have equitable access to clinical studies, regardless of ethnicity, race, or sex.

5. Conclusion

In conclusion, defining and understanding the catchment area for a cancer program allows for subsequent evaluation of clinical trial participants to demonstrate whether or not clinical trial participants appropriately reflect the demographics of the population served by the institution. We hope that these steps will be helpful to National Cancer Institute (NCI) supported organizations that conduct clinical trials, such as SWOG, to set achievable and representative accrual goals for clinical trial studies. The recommendations in this article can serve as a guide toward a better understanding of the cancer research needs of their member site communities and to help set achievable recruitment goals.

Acknowledgments

This work was supported in part by funding from The University of Texas MD Anderson Cancer Center Duncan Family Institute for Cancer Prevention and Risk Assessment, through the Center for Community-Engaged Translational Research. Many thanks to Michael J. Fisch, MD, MPH (Medical Director, Medical Oncology Programs and Genetics, AIM Specialty Health, Chicago, Illinois) for invaluable editorial assistance. We are also grateful to Electra D. Paskett, PhD (Professor of Cancer Research, Ohio State University Comprehensive Cancer Center, Columbus, Ohio) and Marian L. Neuhouser PhD, RD (Full Member and Program Head, Cancer Prevention Program, Fred Hutchinson Cancer Research Center, Seattle, Washington) for their helpful comments on an earlier version of this manuscript.

References

Clayton, J. A., & Tannenbaum, C. (2016). Reporting sex, gender, or both in clinical research? *Journal of the American Medical Association*, *316*(18), 1863–1864.

Griggs, J., Maingi, S., Blinder, V., et al. (2017). American Society of Clinical Oncology position statement: Strategies for reducing cancer health disparities among sexual and gender minority populations. *Journal of Clinical Oncology*, *35*(19), 2203–2208.

Polite, B. N., Adams-Campbell, L. L., Brawley, O. W., et al. (2017). Charting the future of Cancer health disparities research: A position statement from the American Association for Cancer Research, the American Cancer Society, the American Society of Clinical Oncology, and the National Cancer Institute. *Journal of Clinical Oncology*, *35*(26), 3075–3082.

SEER 2014 Statistic, downloaded from https://seer.cancer.gov/faststats/selections.php?#Output on April 25, 2017
Tai, C. G., & Hiatt, R. A. (2017). The population burden of cancer: Research driven by the catchment area of a cancer center. *Epidemiologic Reviews, 39*(1), 108–122.
Texas Cancer Registry. (2020). *Cancer Epidemiology and Surveillance Branch, Texas Department of State Health Services*. Austin: TX. https://dshs.texas.gov/tcr.

Printed in the United States
By Bookmasters